Vienna Circle Institute Yearbook

Institute Vienna Circle, University of Vienna
Vienna Circle Society, Society for the Advancement of Scientific World Conceptions

Volume 30

Editorial Address
Wiener Kreis Gesellschaft
Universitätscampus, Hof 1, Eingang 1.2
Spitalgasse 2-4, A–1090 Wien, Austria
Tel.: +431/4277 46504 (international) or 01/4277 46504 (national)
Email: vcs@univie.ac.at
Homepage: https://vcs.univie.ac.at/
More information about this series at http://www.springer.com/series/6669

Founded by Friedrich Stadler

The Vienna Circle Institute is devoted to the critical advancement of science and philosophy in the broad tradition of the Vienna Circle, as well as to the focusing of cross-disciplinary interest on the history and philosophy of science in a social context. The Institute's peer-reviewed *Yearbooks* will, for the most part, document its activities and provide a forum for the discussion of exact philosophy, logical and empirical investigations, and analysis of language.

Georg Schiemer
Editor

The Legacy of the Vienna Circle

Springer

Editor
Georg Schiemer
Institut für Philosophie
University of Vienna
Wien, Austria

ISSN 0929-6328 ISSN 2215-1818 (electronic)
Vienna Circle Institute Yearbook
ISBN 978-3-031-80567-7 ISBN 978-3-031-80568-4 (eBook)
https://doi.org/10.1007/978-3-031-80568-4

This Springer imprint is published by the registered company Springer Nature Switzerland AG
The registered company address is: Gewerbestrasse 11, 6330 Cham, Switzerland

Editorial—*The Legacy of the Vienna Circle*

The special part of this volume is based on the international conference "*The Vienna Circle—History and Legacy*," which was jointly organized by the Institute Vienna Circle (IVC) and the Vienna Circle Society (VCS) at the University of Vienna in September 25–27, 2023. The motivating idea for the event was to bring together leading scholars to present their work on the history of the Vienna Circle, its members and close allies, as well as on the subsequent reception and general impact of logical empiricism on post-war philosophy and the science studies. The conference included 16 research talks held over the course of three days, with topics ranging from novel historical research of central figures such as Neurath, Carnap, Waismann, Zisel, and Frank to studies of the significance of their work for the sociology of science, the isotype project, and the development of formal semantics (among other topics). Moreover, several of the talks held in Vienna investigated the more general relevance of philosophical and political ideas first developed in the context of the Vienna Circle for contemporary debates on the debunking of science skepticism, the cognitive vs. non-cognitive distinction, and scientific humanism.

The present volume collects nine research articles based on these contributions which investigate different aspects of the historical and philosophy legacy of the Vienna Circle. In her article, Marta Sznajder focuses on the early development of Janina Hosiasson-Lindenbaum's logic of rational degrees of belief and inductive reasoning. Iulian Danut Toader's article critically discusses Carnap's claim that a proper philosophical analysis of quantum mechanics requires a rational reconstruction of the theory. Sebastian Lutz's contribution focuses on Hempel's mature and (self-)critical discussion of the so-called "received view" of scientific theories from the 1960s. Several contributions to the volume study the political implications of logical empiricism: Christian Damböck analyzes a position he labels *noncognitive deliberation*, as developed by the Vienna Circle and especially by Rudolf Carnap. Silke Körber's article focuses on the correspondence and cooperation between Otto Neurath and the British philosopher L. Susan Stebbing on scientific humanism in the 1930s. George Reisch and Adam Tamas Tuboly discuss the conceptual and institutional elements of Philipp Frank's largely forgotten research project on the sociology of science from the mid-1950s. Martin Kusch analyzes two conflicting interpretations (given by Thomas Uebel and by David Bloor) of the place of the sociology of knowledge and epistemic relativism in Logical Empiricism, in particular, in work by Frank and Neurath. The article by Sander Verhaegh, Eugenio Petrovich, and Gregor Bös studies the post-war American reception of Logical Empiricism based on a mention-based bibliometric analysis. Finally, Alexander Linsbichler, Sophie Juliane Veigl, and Benjamin Smart investigate the relations of Popper's philosophy to millennial scientific pluralism.

I would like to thank Friedrich Stadler, my co-organizer of the 2023 conference in Vienna, as well as the speakers for their contributions to such a stimulating event. I am also grateful to Sabine Koch and Zarah Weiss for their administrative support in organizing the event. Concerning the present volume, let me thank the following colleagues: Friedrich Stadler for his helpful editorial advice and assistance, Zarah Weiss for her professional redactorial work, as well as Florian Kolowrat for his assistance during the preparation of the manuscript. Finally, thanks to Bastian Stoppelkamp as the review editor of the Vienna Circle Institute Yearbook series.

Department of Philosophy and Institute Vienna Circle
University of Vienna
Vienna, Austria
September 2024

Georg Schiemer

Contents

Part II General Part

Part III Reviews

Part I
The Vienna Circle—History and Legacy

Chapter 1
Inductive Reasoning and the Logic of Rational Degrees of Belief in Janina Hosiasson-Lindenbaum's Early Work

Marta Sznajder

Abstract The paper traces the origins and early development of Janina Hosiasson-Lindenbaum's logic of rational degrees of belief. I discuss the key points of the publications that resulted from Hosiasson-Lindenbaum's 1926 doctoral dissertation, together with the new material she developed between 1926 and 1934. I argue that while the understanding of inductive reasoning as forming beliefs of various strengths—and hence the concept of degree of belief, or credence—was present in Hosiasson-Lindenbaum's work from its beginning, the idea that the logic of these degrees of belief is the same as the logic of probability was likely introduced later on.

Keywords Janina Hosiasson-Lindenbaum · Subjective probability · Inductive logic · Degrees of belief · Credence · Lvov-Warsaw School

1.1 Introduction

Janina Hosiasson-Lindenbaum's[1] article "On Confirmation" (Hosiasson-Lindenbaum 1940) has a well-recognised place in the history of inductive logic and confirmation theory. Yet, it was only one of a sequence of articles in which Hosiasson for nearly fifteen years explored various aspects of the interaction between probability and belief. Some of the elements found in "On Confirmation" can be traced back to Hosiasson's earliest philosophical works. And while she is a recognised player in the development of probabilistic approaches to epistemology and confirmation theory in the first half of the twentieth century (Galavotti 2008, 2014), her work has so

[1] Henceforth Hosiasson, as the works discussed here were written prior to her name change.

M. Sznajder (✉)
Institute Vienna Circle, University of Vienna, Vienna, Austria
e-mail: marta.sznajder@univie.ac.at

G. Schiemer (ed.), *The Legacy of the Vienna Circle*, Vienna Circle Institute Yearbook 30, https://doi.org/10.1007/978-3-031-80568-4_1

far received only restricted interpretation. This omission is at least in part due to the previous inaccessibility of her earlier work to the wider academic public. Many of Hosiasson's early papers have been published only in Polish and until recently they were not even accessible online. Combined with a surprisingly low interest in Hosiasson's work among Polish historians of philosophy, it meant that so far there has been virtually no extended analysis of the full body of her philosophical work—a situation that affected our understanding of her early period particularly strongly, as in her later years she tended to publish her work in international journals. In order to start bridging this gap, in this paper I take a closer look at some of Hosiasson's earliest work: her doctoral thesis, or rather the papers that came out of it.

Hosiasson received her doctoral degree in December 1926, on the basis of a dissertation titled "Justification of inductive reasoning" ("Uprawnienie rozumowania indukcyjnego").[2] It seems that no copies of the dissertation itself have survived. In the following years, however, Hosiasson published two papers that were based on this work. The first of them, "Definitions of inductive reasoning" (Hosiassonówna 1928), Hosiasson presented at a young philosophers conference in Warsaw in 1927. In the introduction to this paper, she writes that it is based on the first part of her dissertation. The material from the second part of the dissertation was published only in 1934, as an article called "On the Validity of Hypothetical Induction" (Hosiasson 1934), in a volume put together by Tadeusz Kotarbiński's students on the 15th anniversary of his teaching at the University of Warsaw (Geblewicz et al. 1934).

No other published papers of Hosiasson are marked as having originated from the dissertation, but of course that in itself does not mean that there were no further parts of it to begin with. As I discuss below, Hosiasson revised her early approach a great deal in the 1934 publication, and it seems likely that if there were any further chapters in the original text, she might have deemed them obsolete by the time she completed the 1934 revision.

In what follows, I will discuss the key points of the two papers, aiming at developing an understanding of the philosophical project that Hosiasson was initially engaged in, of the decisions she was taking on the way to answering her main philosophical question, and what influences might have played a role in her making those decisions. I am particularly interested in tracking the point at which the notion of probability enters her work, and what subsequently was the role that this notion played in her approach. In my reading, the early Hosiasson turns out to be engaged in the task of developing a theory of rational (degrees of) belief, rather than providing an interpretation to the concept of probability.

---

[2] Janina Hosiasson's student file, University of Warsaw Archive, RP 5224.

1.2 Everyday Inductive Reasoning

"Definitions of Inductive Reasoning" (Hosiassonówna 1928) was the first full paper published by Hosiasson. Based on the first part of her dissertation, its main aim is to clarify the concepts needed to properly state, and then answer, the question of whether inductive reasoning can be justified. In order to formulate the justification question precisely, it is necessary to first define what is actually meant by inductive reasoning. Hosiasson's focus is on what she calls common inductive reasoning, "examples of which we find at every turn in everyday life"[3] (Hosiassonówna 1928, 352).

Of those everyday examples, Hosiasson starts the paper with listing three. First, it can be said of a person who avoids contact with fire after having been burned by it, that she has reasoned inductively. Second, the same can be said of someone who, upon entering their house and seeing that the lock is broken and the rooms empty, assumes that the house had been broken into by thieves who stole what was in the house. Finally, the reasoning performed by an insurance company, which concludes that the probability of death within the next two years of a person of given characteristics is 0,05, based on statistical data showing that of people with the same characteristics, about 5% died in a given number of years, can also be said to be inductive. What these three examples have in common is that they all talk about forming beliefs on the basis of some data—be it experiences, or observations. In none of these motivating examples are explicit generalisations formed.

In the remainder of the paper, Hosiasson wants to provide definitions of inductive reasoning that would be consistent with such an "everyday understanding of this concept" (Hosiassonówna 1928, 352). However, she does not actually start with any supposedly pre-theoretical ideas, instead turning directly to the works of theoreticians of induction, i.e., philosophers whose work she takes to inform the common conception. According to Hosiasson, even the common understanding of what inductive reasoning is, is formed on the basis of the works of theoreticians such as the philosophers she cites, whose results trickle down to school books and the like. It is on this basis that non-philosophers form their understanding of inductive reasoning.

The philosophers whose work Hosiasson based her analysis on were: F. Bacon, J. Herschel, W. Whewell, J. S. Mill, S. Jevons, C. Sigwart, C. S. Peirce, J. Łukasiewicz, K. Twardowski, C. D. Broad, W. E. Johnson, and J. Nicod. This list is not surprising, given Hosiasson's education and philosophical milieu. The inclusion of those members of the Lvov-Warsaw School who have published anything on the topic was an obvious starting point. Then, the strong British presence in the list might have been an influence of Hosiasson's supervisor Kotarbiński, who taught a lot of courses on British empiricism and related topics. Hosiasson took classes with Kotarbiński on Bacon's methodology and Mill's theory of induction as early as her second semester

[3] All translations from the Polish originals in this paper are by me.

at the University of Warsaw.[4] A notable omission on the list is Hume, who Hosiasson never mentions in any of her works, including in the papers on the justification of inductive reasoning. A possible explanation for this omission in this case was that in her choice of philosophers to base her analysis on, Hosiasson was interested in positive accounts of inductive reasoning.

Hosiasson took the basic understanding of what kind of act *reasoning* is from Kazimierz Twardowski, the founder and intellectual father of the Lvov-Warsaw School, whose work was foundational for many of the philosophical developments within the group. In "Definitions", she directly referenced Twardowski's introduction to logic textbook, published in 1901 primarily as a textbook for future teachers and for self-study (Twardowski 1901). There, Twardowski—a former student of Franz Brentano—defines reasoning as "mental acts, in which we conclude that the relation of reason and consequence holds between two or more judgements" (Twardowski 1901, 29). Additionally, Twardowski upheld a strict distinction between reasoning and inference. In acts of reasoning, the relation between reasons and their consequences is established (or rather discovered) regardless of whether the reasons are thought or known to be true. Inference, on the other hand, proceeds from judgements known to be true. Yet, Hosiasson chose not to stick to this latter distinction too closely and did occasionally use the two terms as nearly exchangeable (Hosiassonówna 1928, 353, ft. 3). Nevertheless, this conception of the nature of reasoning did inform Hosiasson's view on the logical form of inductive reasoning. She took inductive reasoning to proceed from some judgements (beliefs) to other judgments (beliefs), on the basis of a consciously recognised reason-consequence relation holding between these judgments.

1.3 Types of Inductive Reasoning

With these preliminaries in place, Hosiasson moves onto analysing the different historical proposals for the specific form of inductive reasoning, or rather the form of its premises and conclusions. She starts with what historically had been the paradigmatic case of inductive reasoning: forming universal generalisations on the basis of singular observations. She calls such reasonings *generalising inductions* (indukcje generalizacyjne): forming the belief that all A's are B based on the fact that a sequence of objects A_1, A_2, etc., which are A, are also B.

There is some tension in the article between honouring the traditional view of inductive reasoning as forming universal generalisations on the basis of particular observations, and wanting to grasp the most general form of inductive reasoning, where the specific logical form of its premises and conclusions is not relevant. In the end, Hosiasson commits herself to the latter aim, and in the sequel to "Definitions" she focuses only on the kinds of inductions where the precise form of the premises

[4] Janina Hosiasson's student file, University of Warsaw Archive, RP 5224.

and conclusions is not relevant. After all, none of the motivating examples she introduced at the beginning of the paper mentioned forming beliefs in the forms of explicit generalisations. Additionally, she pointed out the difficulties associated with trying to spell out precisely the potential additional premises involved in generalising inductions (like the assumption that the world is uniform in some way).

An important feature of such generalising inductions is the fact that their conclusion—"all A's are B"—is a logical reason for (some of) its premises—for instance, that A_1, which is A, is also B. Hosiasson (1928, 357) notes that this feature is precisely what characterises much of the reasoning practiced within sciences that are traditionally called the inductive sciences. Hence, one can define a whole class of reasonings, called *hypothetical inductions* (indukcje hipotetyczne). In hypothetical inductions, we conclude that *b*, on the basis of our judgement that *a*, and that *b* is a (logical) reason[5] for *a*.

Described like this, inductive reasoning looks somewhat like an inversion of deductive reasoning: instead of reasoning from the usual premises to the consequences that follow logically from them, we rather reason from the logical consequences to what entails them. At the same time, Hosiasson saw herself as understanding inductive reasoning in a way different from "the followers of the inversion theory of induction" (1928, 357, ft. 14), as she took them to uphold that inductive reasoning is essentially an act of finding or matching reasons for or to facts. Whereas for her it is an act of changing the degree of belief in consequences on the basis of their reasons—this naturally presupposes that the appropriate reasons have been identified, but goes beyond that. So while in the general outline her conception is an "inversionist" one because it considers induction to proceed from consequences of hypotheses to hypotheses themselves, it is not actually tantamount to the inversion theory in the sense that it focuses not on the act of matching hypotheses with their consequences, but on the subsequent changing of beliefs in them.

Additionally, the inductions are found to also come in versions where we reason not from consequences to their reasons, but from consequences of a hypothesis to other consequences of a hypothesis—these are the *generalising/hypothetical subinductions* (subindukcje generalizacyjne/hipotetyczne). In "Definitions", Hosiasson notes that analogical reasoning belongs to this type. Later on, she developed this analysis into a dedicated theory of inductive reasoning by analogy (see Lindenbaum Hosiasson 1941; Sznajder 2024). However, in the two papers discussed here, she does not yet explore this route; on the contrary, when in the 1934 paper discussed below she suggests a formalisation of hypothetical subinductions for the sake of

[5] Additionally, Hosiasson considers cases where *b* does not logically entail *a*, but *a* is probable given *b*: calls *a* is then an approximate consequence (przybliżone następstwo) of *b* (1928, 361-362). The resulting *approximate inductions* can be found in statistics—the reasoning of the insurance company in one of her initial examples belongs to this category. However, in the remainder of this paper, and in her subsequent work on the logic of induction, she focuses only on the case of logical, and not just probabilistic, consequence relation between the conclusion and the premises of inductive reasoning.

answering the question of their justification, she bases it on a very counterintuitive reading of her description of it.

The aim of "Definitions" was to provide a characterisation of inductive reasoning that would allow for a formulation of the question of its justification. Hosiasson eventually lines up four definitions: of generalising inductions and subinductions, and hypothetical inductions and subinductions. In the present context the hypothetical variants are relevant, and their final definitions can be paraphrased as:

Hypothetical induction is a type of reasoning where we believe, or strengthen our belief, in *b*, on the basis of judgements that *a*, and the belief that "a" is a consequence of "b".

Hypothetical subinduction is a type of reasoning where we believe, or strengthen our belief, that *c*, on the basis of judgments that *a*, and the belief that "a" and "c" are consequences of one and the same thesis "b".[6]

Given these definitions, one can formulate the justification question in its most general form: when we ask whether inductive reasoning is justified, we are asking whether we are allowed—it is rational to do so—to raise our degrees of belief in the above ways. It is that focus on forming beliefs that characterises Hosiasson's approach, and especially the insistence that when reasoning inductively, we might not form full beliefs in the conclusion of the reasoning, but only strengthen a pre-existing belief (Hosiassonówna 1928, 363). But in order to now provide a formal theory of inductive reasoning as defined above—using the notion of the strength of belief—one will need something like a theory of degrees of belief. The natural next step would be to develop this concept of rational degree of belief formally, in order to be able to answer the justification question. The framework eventually proposed in the following paper, "On Validity", to tackle this question is an axiomatic theory of rational degrees of belief.

1.4 Inductive Reasoning and Beliefs in Probabilities

Before the next paper begins, however, "Definitions" ends with a mention of Keynes's view on the form of inductive reasoning. Hosiasson says that according to Keynes, the inductive reasoning concludes not in a change in strength (degree) of belief in the conclusion, but rather in a (full) belief in a statement about a probability value of the conclusion. She raises an issue with this approach:

> If this were the case, the issue of the justification of inductive reasoning would present itself more clearly than under the formulation assumed in the present discussion. For we would not be dealing with the hard-to-capture concept of the degree of belief, and the justification would concern only beliefs about certain values of probability. However, the issue under discussion would gain in clarity only if the concept of probability was clear in the above context. None of the proponents of this view adequately explain in what sense the term

[6]Adapted from (Hosiassonówna 1928, 364), with minor omissions to increase clarity.

> "probability" is used when applied to ordinary inductive reasoning. The issue of clarifying the meaning of the term "probability" seems to me to be very important for the problem of the justification of inductive reasoning even under the view assumed in this article: for it is possible that the "strength of belief" applied to the conclusion of inductive reasoning should depend on the probability of that conclusion (Hosiassonówna 1928, 367).

Hosiasson makes two points here. The first is about the presence of probability in ordinary inductive reasoning. Hosiasson was well-versed in modern statistics, and by 1929 she was working in Jerzy Neyman's laboratory in Warsaw (see, e.g., Hosiasson 1930). From this perspective, the idea of purely epistemic probability, applied to possibly one-off, non-repeatable events, apparently called for a better explanation than the ones provided by Keynes. And the kind of reasoning that Hosiasson focuses on seems to have nothing to do with forming beliefs about such probabilities anyways. Coming back to the initial examples used by Hosiasson: the person who is coming to the conclusion that their house was robbed by burglars, does not seem to be forming beliefs about the probability that their house was broken into, given that the lock is destroyed: they simply believe (to some degree, higher after than before having seen the broken lock) that someone has broken into their house and stole their things. While the proposition that is the object of this belief can be expressed by many statements, the concept of probability is not a part of the content of the belief. A possible exception here could maybe be reasoning involving statistical hypotheses; but even this kind of reasoning is not easily seen as concluding in a full belief about the probability of a hypothesis about probabilities—there appears to be no such stacking of probabilities there.

Hosiasson's second point here is that in spite of not featuring explicitly in premises and conclusions of inductive arguments, probability might still be relevant for degrees of belief, in the sense of being a sort of a normative guide for them. Unfortunately, she does not elaborate on the extent of this connection in "Definitions", nor in the paper that extended it.

1.5 "On Validity": The Logic of Rational Belief

In 1934, Hosiasson published a sequel to "Definitions", titled "On the Validity of Hypothetical Induction" (Hosiasson 1934), which was also partly based on her doctoral work. She seems to have reworked the original text to a large extent. In the beginning of the article, she indicated which parts of it were originally in her dissertation, and which contain new material. She does not list specific changes, but gives a rough description: the first part of "On Validity" contains partially new material, the second part of it is entirely new, and the third part is an elaboration of some parts of the original thesis.

As it turns out, there was a mistake in the printing of the book. Section 2 of Hosiasson's paper was meant to begin on page 28 rather than on page 18. The book copies and individual paper offprints all contained this error, which Hosiasson corrected by hand in the offprints which she distributed to her colleagues. And in the copy gifted to Stanisław Leśniewski, she included a separate errata note, where she explicitly listed this mistake. Knowing this, we get a more correct picture of what definitely was not, and what could have been included already in Hosiasson's 1926 work, because of knowing how much of the content of the 1934 paper was definitely not already present in the dissertation.[7]

The first part of the paper reprises the conclusions of "Definitions", listing again the definitions of the different types of inductions and subinductions, as well as restating the justification problem. The "new material" of this first section consists of a few clarifications of the concepts used in the first paper, like the concept of belief that she uses (degree of belief rather than full belief), or what does it mean to be a logical consequence of a sentence (Hosiasson takes here what is a somewhat beefed-up "inferentialist" approach, where one sentence is said to follow from another whenever it can be derived from it, using the accepted logical theory with all its theorems—*as well as* any other accepted theories, e.g., scientific theories), or, finally, the relativisation of inductive reasoning to the background knowledge of the reasoning agent. The final, third part of the paper, is said to contain some new elements; it appears to be the case that the arguments contained in there might have already been in the original dissertation, but in this paper were restated in the language of the new theory Hosiasson developed in the rewrite.

It is the middle, second part of the paper, that is marked out as being entirely new, and hence not coming from the original doctoral dissertation. This is the material that Hosiasson must have developed between 1926 and 1934. We can only speculate how this happened. After all, quite a long time had passed between her 1926 graduation and the publication of "On Validity"—it is only natural that a philosopher would further develop her views. However, two things happened between 1928 and 1934, which at first glance seem highly relevant for the possible origins of this paper. First, in 1929/1930 Hosiasson spent an academic year in Cambridge, where she became acquainted with the then yet unpublished "Truth and Probability" paper by Frank Ramsey (Ramsey 1931). Second, Stefan Mazurkiewicz, a mathematician in Warsaw, published a set of probability axioms that applied to logical theories (Mazurkiewicz 1932). I will come back to the question of these possible influences below.

The two justification questions posed earlier in "Definitions" were the following. First, the justification of hypothetical induction: (when) are we allowed to believe, or raise our degree of belief in, a hypothesis, on the basis of a consequence of it? Second, the justification of hypothetical subinduction: (when) are we allowed to believe, or raise our degree of belief in, a consequence of a hypothesis, given other consequences of it? In both cases, the question is to be understood as asking when is it rational to change the beliefs in the given way.

---

[7] The offprints of "On Validity" with Hosiasson's handwritten corrections can be found in Warsaw University Library under signatures 121829, 16973, and U.13620 — the last one being Leśniewski's copy with the errata note added in the back.

A natural way to start approaching these questions would be through an analogy with deductive reasoning. Deductive reasoning is considered justified—it is rational to believe its conclusion, given its premises—when its conclusion *in fact* follows logically from its premises. While we cannot require the same of inductive reasoning, maybe we could consider it justified whenever its conclusion partially follows from its premises, i.e., when the premises together with some supplementary statements logically entail the conclusion. Hosiasson (1934, 17) calls this the partial entailment conception, and immediately proceeds to show its triviality: for there would be many cases where any two sentences A and B, could be supplemented by an implication from A to B, in order to form a supposedly justified inductive argument with B as conclusion and A as premise.

Hosiasson (1934, 18) suggests a modification of the partial entailment conception, based on what she took to be one of Keynes's intuitions. According to the second version, we are allowed to believe in the conclusion of an inductive argument ("an induction") to a degree that is the higher, the fewer extra assumptions we would have to supplement it with to make it deductive. Yet again, Hosiasson proceeds to show that even under the most favourable interpretation of the phrase "the fewer extra assumptions" in the above characterisation, the idea is unfeasible.

Hosiasson presented her arguments against the partial entailment conception at the 8th World Philosophy Congress in Prague in 1934 (Hosiasson 1935b, 1936). Somewhat surprisingly, however, the setup of the congress presentation was very different from the one in "On Validity". Unlike the original version, where the motivating question was one of the justification of induction, the congress paper was introduced as one looking for the connection between probability and logical derivability. In the congress paper, which was in German and which Hosiasson must have written herself, she used the word "probability" (*Wahrscheinlichkeit*) and did not mention degrees of belief or their rationality. At the same time, in the Polish original, as well in its English abstract, Hosiasson was careful to not equate the notions of probability and rational degrees of belief, highlighting only the structural equivalence of the formal theories of both concepts.

Given that the Prague presentation and her subsequent talk at the 1935 Unity of Science Congress in Paris (where she discussed the question whether Reichenbach's probability logic was a generalisation of classical logic) were Hosiasson's most high-profile international philosophy lectures, it is no surprise that she was eventually labeled as someone who was interested in the logic of probability, or who pushed for some version of Keynesian logical probability interpretation. The underlying initial interest in the logic of rational degrees of belief, of a much more subjectivist and even psychological nature, got buried under a thick layer of obscurity, due to having been expressed only in Polish and to Polish audiences.

Going back to the 1934 paper, the attempt to formalise the second version of the partial entailment conception—which talks about how strongly we can believe the conclusion of an inductive argument—brings Hosiasson to the topic of formalising the concept of degree of belief. Hosiasson proposes a formal theory of rational degrees of belief, based on Stefan Mazurkiewicz's (1932) axioms for probability functions applied to sentences of a formal language (as opposed to events).

In Hosiasson's own words from the English abstract of that paper, the axioms she proposes are "axioms of >>justified belief<< analogous to axioms of probability" (Hosiasson 1934, English abstract on page 208). For said concept of justified belief, Hosiasson uses the Polish term *wiarygodność*, which can be translated as credibility.

Hosiasson's credibility functions $w(x, Y)$ are functions that apply to pairs consisting of a sentence x and a set of sentences Y: $w(x, Y) = n$ says that the credibility value of x given Y is n. The value of w is the number corresponding to the rational degree of belief in x, given Y, with certainty being assigned value 1. A list of four axioms is proposed "as an expression of the intuitions connected to the concept of credibility" (Hosiasson 1934, 20); further on, Hosiasson characterises the axioms as being of "a rather a priori nature" (1934, 34). The axioms are as following:

Axiom I. If Y entails x, then w(x,Y) = 1.
Axiom II. If Y entails $\neg(x \wedge y)$, then $w(x \vee y, Y) = w(x, Y) + w(y, Y)$.
Axiom III. $w(x \wedge y, Y) = w(x, Y) \times w(y, Y \cup \{x\})$.
Axiom IV. If the sets X and Z are equivalent, then $w(x, X) = w(x, Z)$.

Hosiasson used the same set of axioms in her subsequent work up to the 1940s, including the papers on analogical reasoning (Lindenbaum Hosiasson 1941) and on confirmation (Hosiasson-Lindenbaum 1940), as well as her last work (Sznajder 2024)—the only thing that changed throughout the different iterations of the formal system was the notation and the concept that these axioms were supposed to model (rational degrees of belief, confirmation).

Using these axioms, and the fact that many known theorems of probability theory have a direct translation into her "credibility theory" (1934, 28), Hosiasson proves a number of results: first pertaining to the partial entailment conception of the validity of inductive reasoning, and then, once the latter is shown to be unfeasible, reproving and extending some other results directly related to the two justification questions posed at the beginning of the paper.

The first justification question, following Nicod, is found to have a positive answer. While she seemingly takes the a priori nature of her axioms for rational degrees of belief as self-evident, Hosiasson does not assume nor attempts to prove that induction can be justified on purely logical or priori ground. The substantial assumption needed for inductive reasoning to go through, following Nicod, is what Hosiasson expresses as her formula δ: this is the assumption that only the (logical) consequences of any set Y are certain given Y.

Hosiasson takes herself to be discussing the second justification question in the final, third part of the 1934 paper. However, the way she formalises it turns it into a question of whether the credibility of the "next" consequence of a hypothesis tends towards certainty in the limit, given a growing list of other, already observed consequences of that same hypothesis. Spelling it out like this connects it to similar results discussed by Keynes (1921), and allows Hosiasson to connect her work on inductive reasoning with the topics she worked on as a member of Neyman's team (Hosiasson 1930). In this context, Hosiasson also briefly discusses an alternative to Keynes's assumption of limited variety.

While Hosiasson mentions that there are further important problems to be investigated with respect to the concept of degrees of belief—how to measure actual degrees of belief, and whether we can know any credibility values that are not calculated from other, already known values (similar to what is known in Bayesian epistemology as the problem of the priors)—unfortunately she postpones the discussion of them to future papers, which never materialised.

1.6 Between Warsaw and Cambridge

The main conceptual structure of the two articles that came out of Hosiasson's doctoral dissertation can be summarised as follows. Inductive reasoning is found to be the kind of reasoning where we change the strength of our belief in the conclusion on the basis of some statements that follow from it. To ask when such reasoning is justified, is to ask when such a belief change is rational. To answer the latter, we look at rational degrees of belief: inductive reasoning is justified when the rational degree of belief in its conclusion, given its premises, is in fact higher than the rational degree of belief in the conclusion, without the premises[8]—which can be proved within the presented axiomatic theory of credibility (with some additional assumptions). The axioms themselves are analogues of axioms for probability functions of Mazurkiewicz.

In the first of the papers, there are no hints towards the possibility of the theory of probability being of any use in constructing the theory of rational degrees of belief. In the final remark on Keynes, discussed above, Hosiasson acknowledges that the specific values for the rational degree of belief in the conclusions of an inductive argument could depend on their actual probability—but that in itself does not entail that the formal theories of those two concepts would be essentially identical.

As a matter of fact, Hosiasson continued to express an agnostic attitude towards the relation between the concepts of credibility and probability. At first, she described her axioms only as being analogous to the axioms of probability; no further connection was made between the two concepts themselves. In subsequent papers, she maintained this kind of cautious attitude towards this issue. For instance, when introducing her axioms in the 1941 paper on analogical reasoning, she once again refrained from taking a stance: "I do not want to decide here whether (partial) certainty and probability are identical concepts" (Lindenbaum Hosiasson 1941, 354, translation mine).

Consequently, the conceptual setup by the end of "Definitions" raises a number of questions. The main thesis of that paper was that inductive reasoning is a process in which the strength of belief in certain statements changes. How could one proceed onwards from this point: what kind of further theory would be needed to start

[8] To be exact, credibility here is always relativized to *some* statements: the background knowledge.

answering the justification question as formulated in the paper? One would expect some sort of theory of rational degrees of belief—and this is exactly what she delivered in the "new material" section of the following paper.

But what could such a theory look like in 1926, when Hosiasson was writing the original doctoral thesis? What conceptual options did she have in 1926 to tackle this question? The answers to these questions are by necessity purely speculative; there are none of Hosiasson's notes, drafts, or even relevant correspondence available from this period. As I have mentioned earlier, two events appear relevant to the development of the new approach in "On Validity": Hosiasson's stay in Cambridge and her reading of Ramsey's "Truth and Probability", and the publication Mazurkiewicz's work on probability functions applied to formal languages. However, Hosiasson explicitly mentions only the second of these texts in her paper, i.e., Mazurkiewicz's axiom system. No reference is made to Ramsey's work, in contrast to other papers of Hosiasson (1931, 1935a); and the exact relation between the work of these two philosophers is a topic deserving a separate study.

The importance of Mazurkiewicz's work for the development of Hosiasson's theory of inductive reasoning is clear: it was the first formal theory of probability that she could have access to, in which probability was applied to sentences, which is exactly what she needed for her credibility functions. In statistics, which she practiced under Neyman, probabilities apply to events, not sentences of a given formal language. The traditional setup was not applicable to the kind of thing she wanted to model. Hence, it is hard to imagine any proto-version of the 1934 system that could already be present in the original text from 1926.

The third part of the 1934 paper, marked as containing material that was already present in the dissertation, suggests that in 1926 Hosiasson did work within Keynes's own framework, in spite of her distancing herself from some fundamentals of his approach. The main source of theorems that Hosiasson reproves in her paper is Keynes's *Treatise*. Hence, this early work also provides some more context to the extent of Keynes's influence on Hosiasson. It is true that she was reading his work on probability in detail, and took a lot of inspiration from his suggestions and results for her own work. But this influence rather takes the form of her taking some of his formal problems, or theorems, and working on improving them (or disproving them, as in the case of the idea that the strength of inductive arguments is somehow related to the degree of partial entailment). It does not seem to be the case that she took up his overall philosophical approach wholesale. At most she stayed agnostic as to the nature of probability that Keynes was suggesting; and at times she was explicitly skeptical.

Finally, Hosiasson's focus in the early works is purely epistemic: inductive reasoning is a way of forming beliefs, and there is no discussion on how those beliefs interact with decisions or action (with the one exception in the very beginning of the first paper, where the first example of everyday inductive reasoning is of a person who avoids direct contact with fire after having been burned before). Hosiasson never attempted to provide any justification for her axioms for credibility functions, or any justification for them being identical to probability axioms—whether in those early papers, or later on.

This kind of purely axiomatic approach to philosophy could be traced back to the influence of Jan Łukasiewicz, who was one of Hosiasson's teachers at the University of Warsaw and a co-supervisor of her PhD thesis. Łukasiewicz explicitly formulated the ideal that philosophical work should follow the axiomatic, formal ideal of mathematical logic and that the axioms of philosophical theories should be intuitively clear (see Brożek 2022). Hosiasson's approach in "On Validity" fulfils these exact desiderata.

1.7 Conclusion

Janina Hosiasson-Lindenbaum is most known for the axiom system for confirmation functions and her use of it to provide a response to Hempel's raven paradox, which she put forward in her 1940 paper "On Confirmation". While she did not elaborate then on the precise meaning given to the functions, the flavour of that proposal was quite distinctly objectivistic. It is not surprising that Rudolf Carnap took this work to be an important predecessor of his own inductive logic, which started off with a focus on an interpretation of confirmation as an objective probability concept (Carnap 1950, §62).

But it turns out that Hosiasson had developed that axiom system already in the early 1930s, albeit for quite a different purpose. In this earliest iteration, Hosiasson's axiom system was meant to be a theory of rational degrees of belief, i.e., credibility functions, not at all spelled out in terms of objective meanings of any sort. This was a different approach than an inductive logic that is supposed to capture some objective relation holding between propositions, like in Keynes's theory—even if, as acknowledged by Hosiasson, rational degrees of belief might be somehow related to objective probabilities in the end. In that sense, these early works of Hosiasson place her very close to the way Carnap thought of his own inductive logic project in its late stages, after he embraced the interpretation of his logical probability functions as credibility functions.

But unlike Carnap, Hosiasson does not see herself as providing an interpretation to the probability concept (or even one of the two existing probability concepts), at least not in the early works that stemmed from her doctoral thesis. Her initial interest was epistemological: she wanted to develop a formal theory of ordinary inductive reasoning. And while a lot of her formal results were conceived of in dialogue with Keynes, philosophically her approach seems to have stemmed primarily from the methodological ethos of the Lvov-Warsaw School, directly through her teachers in Warsaw, as well as indirectly through Twardowski's work.

References

Brożek, A. 2022. Jan Łukasiewicz's Program of the Logicization of Philosophy: Its Genesis, Content and Realizations. *Synthese* 200: 199.

Carnap, R. 1950. *Logical Foundations of Probability*. Chicago: University of Chicago Press.

Galavotti, M.C. 2008. A tribute to Janina Hosiasson Lindenbaum. A Philosopher Victim of the Holocaust. In *The Migration of Ideas*, ed. R. Scazzieri and R. Simili, 179–194. Watson Publishing International LLC.

———. 2014. Probabilistic Epistemology: A European tradition. In *European Philosophy of Science – Philosophy of Science in Europe and the Viennese Heritage*, Vienna Circle Institute Yearbook, ed. M.C. Galavotti, E. Nemeth, and F. Stadler, vol. 17, 77–88. Cham: Springer.

Geblewicz, E. et al., eds. 1934. *Fragmenty filozoficzne. Księga Pamiątkowa ku uczczeniu piętnastolecia pracy nauczycielskiej w Uniwersytecie Warszawskim profesora Tadeusza Kotarbińskiego*. Warszawa 1934.

Hosiasson, J. 1930. Quelques remarques sur le dépendance des probabilités a posteriori de celles a priori (Some Remarks on the Dependence of Probabilities a Posteriori on Those a Priori). In *Comptes-rendus du I Congrès des Mathématiciens des Payes Slaves, Warszawa 1929*: 375–382.

———. 1931. Why Do We Prefer Probabilities Relative to Many Data? *Mind* 40 (157): 23–36.

———. 1934. O prawomocności indukcji hipotetycznej (On the Validity of Hypothetical Induction). In: Geblewicz et al. (1934), 11–34.

———. 1935a. Uwagi w sprawie psychologji wnioskowań indukcyjnych (Remarks on the Psychology of Inductive Reasoning). *Księga pamiątkowa ku czci Władysława Witwickiego*. Special issue of *Kwartalnik Psychologiczny* 7: 275–300.

———. 1935b. Wahrscheinlichkeit und Schluss aus Teilprämissen (Probability and Inference from Partial Premises). *Erkenntnis* 5: 44–45.

———. 1936. Wahrscheinlichkeit und Schluss aus Teilprämissen (Probability and Inference from Partial Premises). In *Actes du Huitième Congrès International de Philosophie à Prague, 2–7 September 1934*, 192–196. Orbis: Prague.

Hosiasson-Lindenbaum, J. 1940. On Confirmation. *The Journal of Symbolic Logic* 5 (4): 133–148.

Hosiassonówna, J. 1928. Definicje rozumowania indukcyjnego (Definitions of Inductive Reasoning). *Przegląd Filozoficzny* 31 (4): 352–367.

Keynes, J.M. 1921. *A Treatise on Probability*. London: Macmillan.

Lindenbaum Hosiasson, J. 1941. Induction et analogie: Comparaison de leur fondement (Induction and Analogy: A Comparison of Their Foundations). *Mind* 50 (200): 351–365.

Mazurkiewicz, S. 1932. Zur Axiomatik der Wahrscheinlichkeitsrechnung (On the Axiomatisation of the Probability Calculus). *Comptes rendus des séances de la Société des Sciences et des Lettres de Varsovie* XXV (1–6): 1–4.

Ramsey, F. 1931. Truth and Probability. In *The Foundations of Mathematics and Other Logical Essays*, with a preface by G.E. Moore, ed. R.B. Braithwaite, 156–198. London: Kegan Paul, Trench, Trubner, & Co., 1931.

Sznajder, M. 2024. Janina Hosiasson-Lindenbaum on Analogical Reasoning: New Sources. *Erkenntnis* 89: 1349–1365.

Twardowski, K. 1901. *Zasadnicze pojęcia dydaktyki i logiki* (The Essential Notions of Didactics and Logic). Lwów: Nakładem Towarzystwa Pedagogicznego.

Chapter 2
Quantum Mechanics as a Carnapian Language

Iulian Danut Toader

Abstract The paper discusses Carnap's claim that a proper philosophical analysis of quantum mechanics, including a determination of whether its logic has to be revised, requires a rational reconstruction of the theory. Several articulations of the notion of rational reconstruction are recalled, followed by a brief analysis of two standard criticisms of Carnap's claim. The paper suggests that adopting inferentialism overcomes both criticisms, and then considers the possibility of formulating quantum mechanics as a Carnapian language with an inferentialist semantics.

Keywords Quantum mechanics · Quantum logic · Rational reconstruction · Metasemantics · Representationalism · Inferentialism

2.1 Introduction

It is an experimental fact that quantum mechanics (QM) makes adequate empirical claims, and we can know that QM makes adequate empirical claims, since there is plenty of inductive evidence supporting this fact. But suppose that a rational reconstruction of QM can be given. A rational reconstruction, as Carnap stipulated, would formulate QM in terms of a partially interpreted formal language that includes, along with the logico-mathematical axioms, theoretical sentences and correspondence rules that partially endow the language with empirical meaning by relating theoretical sentences to observation sentences. However, due to Gödel's second incompleteness theorem, without allowing stronger theoretical resources, the consistency of a rational reconstruction of QM is unprovable. But without a proof of consistency, we cannot know that QM makes adequate empirical claims. Thus, our supposition must be denied: *QM cannot be rationally reconstructed.*

I. D. Toader (✉)
Institute Vienna Circle, University of Vienna, Vienna, Austria
e-mail: iulian.danut.toader@univie.ac.at

G. Schiemer (ed.), *The Legacy of the Vienna Circle*, Vienna Circle Institute Yearbook 30, https://doi.org/10.1007/978-3-031-80568-4_2

One way in which this argument, inspired by Michael Potter's criticism of Carnap's philosophy of arithmetic (Potter 2000), can be resisted is by noting an equivocation between two types of knowledge: *fallible* knowledge, based on inductive evidence, and *infallible* knowledge, based on mathematical proof. There is no reason to dismiss the supposition that QM can be rationally reconstructed, if we can fallibly know that QM makes adequate empirical claims, but at the same time we cannot infallibly know that QM makes such claims. This argument, thus, fails to establish that a rational reconstruction of QM cannot be given.

But are there any good reasons to think that a rational reconstruction of QM *can* be given? And if so, what would such a reconstruction be good for? More specifically, what did Carnap think that a rational reconstruction of QM would be needed for? To address these questions, I will proceed as follows. In Sect. 2.2, I will describe Carnap's reflections on QM, and I will emphasize his claim that a proper philosophical analysis of QM, including a determination of whether its logic has to be revised, requires a rational reconstruction of the theory. After I recall the successive articulations of the notion of rational reconstruction in the development of Carnap's thought, I will turn in Sect. 2.3 to a discussion of whether a rational reconstruction of QM is possible, and I will briefly consider two standard criticisms, recently echoed by Richard Healey (2017) and David Wallace (2020). But I will argue that both criticisms can be met, for they both assume that rational reconstruction necessarily requires representationalism. Adopting inferentialism instead would allow for the possibility of formulating QM as a Carnapian language. In Sect. 2.4, I will briefly revisit Sellars' critique of Carnap, and I will point out that the latter could have indeed adopted inferentialism for non-logical terms, just as he had done for logical terms. I will conclude the paper by clarifying a seemingly worrying circularity problem for my argument, and noting a further line of research, already suggested by Carnap, which is reopened by the possibility claim I defend in this paper.

2.2 Carnap on Quantum Mechanics

Carnap's chapter, "Indeterminism in Quantum Physics", of his 1966 book, starts by presenting some basic principles of QM. Heisenberg's Uncertainty Principle is characterized as "a fundamental law that must hold as long as the laws of quantum theory are maintained in their present form" (Carnap 1966, 284). Carnap duly noted that the limitations entailed by this principle cannot be reduced through any possible improvements of our measuring techniques, since they are not due to the imperfections of our measuring instruments. The mathematical representation of a quantum state by means of a wave function defined on an abstract higher-dimensional space, i.e. on configuration space, is carefully presented. Carnap described the deterministic dynamics of quantum-mechanical systems, governed by the Schrödinger equation, and the probabilistic character of all predictions of the results of any measurements performed on such systems, briefly touching on the QM of macroscopic objects (like satellites) as well. On the basis of his understanding of the basic principles of QM, Carnap offered, throughout the book, his answers to some important philosophical questions, which he obviously thought could already be addressed on that basis.

For example, on Carnap's view, the necessarily statistical character of quantum-mechanical explanations cannot be regarded as a manifestation of our ignorance, but as an expression of the basic structure of the world, which entails that all physical explanations can only be statistical, under the assumption that all laws of physics reduce to the fundamental principles of QM (such as the Uncertainty Principle) (Carnap 1966, 9). Relatedly, Carnap reported as an "interesting speculation" that QM might indicate that this very structure, and thus presumably its fundamental ontology, including space and time, is all discrete, rather than continuous (Carnap 1966, 89). Furthermore, according to Carnap, QM clearly suggests that there can be no explicit definitions of quantum-theoretical concepts in terms of empirical concepts. But he noted that a more satisfactory answer to the question about the empirical meaning of quantum properties like spin, for instance, would require "an elaborate theory" (Carnap 1966, 221). Carnap also maintained that QM is irrelevant to philosophical debates on the existence of free will. This is because, on his view, indeterminate quantum jumps, though random, cannot play any role in decision making since "it is not likely that these are points at which human decisions are made" (Carnap 1966, 221). But even if they were, that would only make our decisions equally random, and so they would simply not be choices at all, but chances. And even if the range of quantum randomness were much greater than in the actual world, as described by QM, that would only decrease the possibility of free choices.

However, according to Carnap, there are also questions that cannot be addressed properly on the basis of the formulations of QM that Carnap was aware of. These include questions about the logic and language of the theory. He wrote: "The revolutionary nature of the Heisenberg uncertainty principle has led some philosophers and physicists to suggest that certain basic changes be made in the language of physics. [...] The most extreme proposals for such modification concern a change in the form of logic used in physics" (Carnap 1966, 288). Among these proposals, he recalled Birkhoff and von Neumann's (1936) change of the transformation rules, by the replacement of the law of distribution (of conjunction over disjunction) by that of modularity. However, unlike Putnam (1968), Carnap was not ready to take lessons in logic from QM. Rather, he was inclined to think that it was not a change in logic, but the change in the causality structure implied by quantum-mechanical laws (i.e. indeterminism), that made QM revolutionary.

The question whether the logic of physics ought to be revised is not one that could be addressed properly without first presenting "the entire field of physics stated in a systematic form that would include formal logic" (Carnap 1966, 290). Even though he is not fully explicit about it, it is quite clear that Carnap demanded a rational reconstruction of modern physics. But he thought that such a rational reconstruction of modern physics had not been given: "[Its] language is still, except for its mathematical part, largely a natural language; that is, its rules are learned implicitly in practice and seldom formulated explicitly" (Carnap 1966, 291). This tacitly implies that even von Neumann's (1932) and Mackey's (1963) rigorous formulations of QM—widely regarded as mathematical axiomatizations of QM *par excellence*—fail to qualify as rational reconstructions of modern physics, despite their mathematical clarity and their transparent use of the axiomatic method. I will come back to this implication in the next section (for some more details on Carnap's view on QM, see Horvat and Toader 2023).

In any case, note that Carnap did not refer to the lack of systematization of *QM in particular*, but that of *modern physics as a whole*: thus, a rational reconstruction of QM alone (without gravitational and/or spacetime physics) would not provide Carnap with a proper basis for discussing alternative logico-linguistic frameworks. Therefore, it seems fair to say that Carnap's unwillingness to take lessons in logic from modern physics is caused not only by the absence of a fully formalized axiomatization of the latter, but also by its disunity, which is still manifest today in the tensions between quantum and gravitational physics. His main argument appears to have been as follows: *One can establish that logic needs to be revised in QM only after modern physics as a whole has been rationally reconstructed. But there is no such rational reconstruction, thus the need to revise logic in QM cannot yet be established.* Some may consider this argument, and in particular the very strong conditions on logic revision, rather preposterous. But I will consider the following watered-down version of it: *One can establish that logic needs to be revised in QM only after QM has been rationally reconstructed. But there is no such rational reconstruction, thus the need to revise logic in QM cannot be established.* Of course, we still have no rational reconstruction of QM, in the sense Carnap demanded. But the main question that I am concerned with is whether such a reconstruction would be even possible.

2.3 Is a Rational Reconstruction of QM Possible?

At this point, let me very briefly recall some of the history of the notion of rational reconstruction in Carnap's works (for more historical details, see Beaney 2013). Although Carnap mentioned this notion for the very first time in his 1928 book, *Der logische Aufbau der Welt*, its roots are of course in Hilbert's views on formal axiomatics. In a famous 1917 lecture in Zurich, Hilbert said: "When we assemble the facts of a definite, more-or-less comprehensive field of knowledge, we soon notice that these facts are capable of being ordered. This ordering always comes about with the help of a certain *framework of concepts…* [which] is nothing other than the *theory* of the field of knowledge" (Hilbert 1918, §2). A Hilbertian framework of concepts is, however, according to the early Carnap, not yet a rational reconstruction: "A theory is *axiomatized* when all statements of the theory are arranged in the form of a deductive system whose basis is formed by the axioms, and when all concepts of the theory are arranged in the form of *a constructional system* whose basis is formed by the fundamental concepts" (Carnap 1928, §2). Thus, a theory is rationally reconstructed only if its framework of concepts is (presented as) a constructional system.

A constructional system is a system of concepts that results from the application of what Susan Stebbing called "directional analysis". Although I do not have the space here to discuss this in detail, here is how she articulated this notion in the case of the system of *Principia Mathematica*: "[This] is not a postulational system. Whitehead and Russell explicitly state that their work 'aims at effecting the greatest possible analysis of the ideas with which it deals.' These 'ideas' are the fundamental concepts of mathematics, which they sought to reduce to their simplest elements. The authors were not concerned to construct a postulational system; they did not seek to obtain *one* out of a set of different postulational systems *any* one of which would yield the required demonstrations regarding a specific set of mathematical statements. They sought a *single* system such that its primitive concepts and its primitive propositions should yield the whole of mathematics. Thus their system is to be uniquely based on a single set of fundamental concepts and postulates; further, it is to admit only of *one* interpretation. Their aim could be achieved if, *and only if*, the primitive postulates, and primitive concepts, are not merely *postulated*, and taken as *undefined*, but are also *fundamental* in a sense which excludes arbitrary selection. Thus the analysis employed is directional" (Stebbing 1932, 90sq). Arguably, the same holds also for Carnap's constructional system, which he presented in his *Aufbau* (for details, see Toader 2015). If this is correct, then a rational reconstruction is a constructional system in Stebbing's sense, i.e., a system of fundamental, non-arbitrarily selected concepts.

After 1934, however, Carnap's view is that the rational reconstruction of a theory is supposed to use the tools developed in his book *Logische Syntax der Sprache*. More exactly, an empirical theory is rationally reconstructed if presented as a formal language together with a set of semantic or correspondence rules that partially connect this language to its empirical interpretations (Carnap 1939, 60). Rational reconstruction thus requires a reformulation of the empirical theory, a reformulation that rigorously defines a consequence relation, and distinguishes logical and descriptive terms on that basis. As is well known, Carnap's view evolved from this so-called partial interpretation approach, to one that would employ Ramsey sentences and, later, the Hilbert's epsilon operator (see, e.g. Demopoulos 2007; Gratzl and Schiemer 2016).

To come back to my questions then: can there be a rational reconstruction of QM, as a constructional system of fundamental, non-arbitrary concepts or as a partially interpreted formal language? What would recommend any of such reconstructions as a proper condition for an evaluation of logic in QM?

What might recommend rational reconstruction as a proper condition for an evaluation of logic in general, for that matter? In 1943, Carnap maintained that a rational reconstruction of classical logic is needed for its metasemantic evaluation, for the rational reconstruction can explicate properties of classical logic such as whether the meaning of its connectives and quantifiers is fixed. This property is explicated by investigating whether the rational reconstruction of classical logic is complete or *categorical*, i.e., whether it has a unique semantics up to a isomorphism. Thus, on his view, a rational reconstruction is needed for understanding the metasemantics of classical logic. Carnap argued that the rational reconstruction he considered, i.e., a

Hilbert-style formal axiomatization of classical logic, is *not* categorical, by constructing what he called non-normal valuations for negation and disjunction, as well as for the quantifiers (Carnap 1943). This raised what came to be called a categoricity problem: same formal rules, different meanings. Importantly, this problem was revealed by investigating a rational reconstruction of classical logic, and led Carnap to propose a revision of its calculus by allowing arguments that have multiple conclusions.

But what would a rational reconstruction of QM be good for? Would such a rational reconstruction be also required for a metasemantic evaluation of QM? If so, then investigating whether the rational reconstruction is categorical might likewise suggest a revision of the QM formalism, perhaps even a revision of its logical rules.

One way to think about this investigation, I suggest, is as a metatheoretical test for quantum inferentialism. As just mentioned, Carnap applied this test to classical logic, but the same test could of course be applied to other theories, such as Peano Arithmetic: Can PA be rationally reconstructed? If it can, then the properties of the reconstruction can explicate some properties of PA. But is the reconstruction categorical? If not, then the arithmetical rules do not fix the meaning of arithmetical terms. Thus, arithmetical inferentialism is false. Suppose we want to run the test on QM: Can QM be rationally reconstructed? If yes, then the properties of the reconstruction can explicate some properties of QM. But is the reconstruction categorical? If not, then the quantum rules do not fix the meaning of QM terms. Thus, quantum inferentialism is false.

The problem with this latter test is, again, that we don't know whether a rational reconstruction of QM is possible. Might standard QM already count as such? Standard QM, i.e., the axiomatization given by von Neumann in 1932, might look like a Hilbertian framework of concepts. But is it a Carnapian constructional system? For one thing, it's not clear that its basic concepts (like the concept of Hilbert space) are fundamental and non-arbitrarily selected, and there is no systematic construction of concepts, as in Carnap's *Aufbau*, either. The standard axioms do not seem to be a result of the kind of directional analysis illustrated in *Principia Mathematica*. Certainly, they do not "admit only of *one* interpretation", as Russell and Whitehead had required for their axiomatization of mathematics. Quite the opposite, in fact, although this depends on what one takes an interpretation to be.

Recent reaxiomatizations and reconstructions of QM (e.g., Hardy 2001) could be considered results of directional analysis, since they purport to derive standard QM from general information-theoretical principles, considered to be more fundamental and less arbitrary than von Neumann's axioms. But these reaxiomatizations are not formulated as Carnapian languages, and they do not even consider the central task of defining a consequence relation for the language of the reconstructions. Perhaps this is not surprising after all: maybe there is no rational reconstruction of QM because there simply cannot be any such reconstruction, in any of the senses of this notion articulated by Carnap.

This conclusion might be implied, for instance, by Healey's rejection of the Carnapian reconstructionist project: "For Carnap, the task of the philosopher is to seek clarification of what a scientific theory says by means of a logical reconstruction of the theory within a precisely defined language. … there is now a consensus among philosophers of science that the labor involved in re-expressing a significant theory in a formal language and then giving its semantics by means of correspondence rules would make this neither a practicable nor a useful technique for revealing its structure and function" (Healey 2017, 123). Interestingly, however, the reported consensus is taken to justify a rejection of rational reconstruction as neither practicable nor useful. But note that Healey does not explicitly maintain it to be an impossible technique. More importantly, note that a rational reconstruction, as in Carnap's partial interpretation approach to scientific theories, is understood to require representationalism, i.e., that the semantics of the formal language of a rational reconstruction is supposed to be given by correspondence rules. This leaves it open that a rational reconstruction were possible, and might be even a practicable and useful technique, if it should not require representationalism, but inferentialism, i.e., that the semantics of the formal language would be given by inferential rules. Healey, as an inferentialist, would likely embrace this suggestion.

However, others reject Carnap's project precisely because they understand it to require non-representationalism, at least for theoretical terms, which is what Carnap had arguably suggested already in his 1927 paper, *Eigentliche und uneigentliche Begriffe*. Criticizing Healey's own inferentialist approach to standard QM, Wallace writes: "The non-representationalist strategy … is not new, nor is it specific to quantum theory. It is, rather, the central idea in the logical-positivist and logical-empiricist pictures of science. … It is almost universally accepted today that these approaches are not viable. But the predominant reason, historically, that they fell from grace was … the increasingly clear realization – notably … in the recognized failure of Carnap's project in the *Aufbau* (1928) – that observation is theory-laden" (Wallace 2020, 92). The predominant reason reported here is implied to justify the rejection of rational reconstruction on the basis of a problematic division between observational terms (admitting a representationalist semantics) and theoretical terms (having a non-representationalist semantics). Since no such division is actually defensible, the failure of both Carnap's reconstructionist approach and Healey's inferentialist approach is unavoidable. But this criticism obviously overlooks the very same possibility that I just suggested above, that the rational reconstruction of a theory should dispense with representationalism *not only for its theoretical terms*. If inferentialism were instead globally required, for all the terms in a language, then perhaps the problematic division between observational and theoretical terms would not be enough for the rejection of rational reconstruction.

But would it be possible for the Carnapian interested in rationally reconstructing QM to adopt global inferentialism, for both theoretical and observational terms in a language? To address this question, even if only succinctly, I propose that we revisit Sellars's inferentialist critique of Carnap's view in the *Syntax*. My purpose in the next section is to suggest a positive answer to this question, rather than rejecting that critique and defending Carnap's view against it (for one such defense, see e.g. Carus 2004).

2.4 Carnap's *Syntax* and Sellars' Inferentialist Critique

Carnap's *Syntax* was the inspiration and, at the same time, the target of Sellars' paper "Inference and Meaning" (Sellars 1953): "Sellars identifies his 'material rules of inference' with Carnap's 'P-rules'. ... Carnap's views ... made the scales fall from Sellars's eyes" (Brandom 2015, 43sq). Nonetheless, against Carnap's view that P-rules are just thought-economical devices, "a matter of convention and hence, at most, a question of expedience" (Carnap 1934, 180), Sellars argued that material rules of inference are indispensable, and irreducible to logical rules. In order to explain his argument, consider an example of a logical inference: *If an object is red, then it's colored. This object is red. Thus, it's colored.* And another: *If something is gray, then it's a slithy tove. Findus is gray. Thus, it's a slithy tove.* What is characteristic of such inferences is, of course, that one can assent to their conclusions once one has assented to their premises, even though one may not know what the terms mean. In a slogan, it's the logical form that matters.

Sellars' slogan was, however, different: it's the matter that matters, not the logical form. In the case of material inferences, he contended, unless one assents to their conclusions once one has assented to their premises, one does not know what the terms involved mean. Here is an example of what he considered material inferences: *This object is red. Thus, it's colored.* And another: *Findus is on the mat. Thus, Findus is not on the roof.* Sellars noted that material inferences are expressed by subjunctive conditionals, such as 'If this object were red, then it would be colored.' and 'If Findus were on the mat, Findus would not be on the roof.' But subjunctive conditionals, as Brandom later emphasized, are in fact implicit modal statements: 'Necessarily, if this object is red, then it is colored.' and 'Necessarily, if Findus is on the mat, then Findus is not on the roof.' Due to this implicit modality, Sellars concluded that (the rules of) material inferences are irreducible to (the rules of) logical inferences. This is so, on his view, because one cannot detach the consequent of a subjunctive conditional *only* by affirming its antecedent. For better or worse, Sellars appears to have taken this argument to establish the metasemantic indispensability of (the rules of) material inferences: such inferences must be regarded as relations between meanings, not as relations between extensions of concepts. Expressing material inferences as implicit modal statements makes this point explicit. He also thought that this corrected Carnap's view.

According to Sellars, Carnap had maintained that the material rules are, in principle, reducible to logical rules: "*Carnap, however, makes it clear that in his opinion a language containing descriptive terms need not be governed by extra-logical transformation rules*. Indeed, he commits himself (p. 180) to the view that for every language with P-rules, a language with L-rules only can be constructed in which everything sayable in the former can be said" (Sellars 1953, 320; original emphasis). But it is rather hard to endorse this reading of what Carnap actually says on page 180 of his *Syntax*, and one would have to say much more to justify the step from P-rules being a matter of convention and expedience, as Carnap saw them, to P-rules being metasemantically dispensable, in the sense Sellars thought Carnap saw them. On Sellars's reading, Carnap regarded material rules as admissible only on account of "the economy in the number of premises required for inferences" (Sellars 1953, 321). So Sellars considered Carnap's view to be that material rules are inessential to any language, and thus metasemantically inert. What Carnap actually maintained is that, in principle, one could stipulate only logical rules or one can adopt as material rules all sentences that are not logical rules: the choice is based on pragmatic criteria like simplicity and fruitfulness. He never said that adding or dropping P-rules would leave the expressive power of a language unchanged. In any case, Sellars doubted that the view that material rules are metasemantically dispensable is correct: "But might it not be possible for an empiricist to hold that material rules of inference are as essential to meaning as [the logical] rules? … P-rules are essential to any language which contains non-logical or descriptive terms" (Sellars 1953, 336). Indeed, Sellars believed that his argument from the implicit modal character of material inferences successfully refutes what he took to be Carnap's view.

But Carnap's reflections on P-rules actually raise an important point for Sellars' inferentialism, and furthermore emphasize that the two views are actually compatible. Carnap wrote: "If P-rules are stated, we may frequently be placed in the position of having to alter the language; and if we go so far as to adopt all acknowledged sentences as valid, then we must be continuously expanding it. But there are no fundamental [as opposed to practical] objections to this" (Carnap 1934, 180). Continuous expansion of a language and continuous modification of its semantics, although impractical, may be admissible. But on Sellars' inferentialism, such a modification would be unavoidable. The introduction of new vocabulary, and as a special case the introduction of new concepts in science, always allows us to make novel material inferences. Given the metasemantic work done by such inferences, on Sellars' view, they will not only determine the meaning of the new vocabulary, but will change the meaning of at least some of the old expressions as well. This updating of semantics presupposes, in any case, that material rules are such that novel material inferences can be made when new concepts are introduced in a language.

Moreover, as Carnap rightly noted, rules can also be altered: "No rule of the physical language is definitive; all rules are laid down with the reservation that they may be altered as soon as it seems expedient to do so. This applies not only to the P-rules but also to the L-rules, including those of mathematics. In this respect, there are only differences in degree; certain rules are more difficult to renounce than others" (Carnap 1934, 318). Thus, for Carnap, updating the semantics of a language as a result of changing the rules of inference would be admissible as well, while for Sellars this updating would be, once again, unavoidable. If material inferences determine the meaning of empirical terms, every change of rules implies a modification of the meaning of at least some of these terms in the language.

Thus, whereas for Carnap, the continuous expansion of an empirical language, the modification of its rules, and the updating of its semantics, are all admissible, for Sellars they are unavoidable. This suggests that if good reasons had become apparent to him, Carnap would not have dismissed an inferentialist semantics for empirical vocabulary. Carnap adopted inferentialism with respect to logical terms, as can be seen most clearly in his preface to the *Syntax*: "Let any postulates and any rules of inference be chosen arbitrarily; then this choice, whatever it may be, will determine what meaning is to be assigned to the fundamental logical symbols" (Carnap 1934, xv, emphasis removed). But if my observations above are correct, then it seems clear that Carnap could have embraced inferentialism, not only with respect to logical terms, but with respect to non-logical, empirical terms as well, for there was no principled reason for him to resist it.

2.5 Conclusion

My argument, as presented so far, seems to have a circularity problem: for it first stipulates that testing inferentialism requires a rational reconstruction of QM, but then emphasizes that a viable rational reconstruction of QM would better drop representationalism and adopt, instead, global inferentialism. However, there really is no circularity, for the argument actually says that testing inferentialism *for the language of QM* requires a rational reconstruction of QM, but a viable rational reconstruction of QM requires adopting inferentialism *for the language of that reconstruction*. There is no circularity involved in a test of inferentialism for the language of QM that requires adopting inferentialism for the language of the reconstruction.

In 1966, Carnap issued a call "for close cooperation between physicists and logicians—better still, for the work of younger men who have studied both physics and logic. The application of modern logic and the axiomatic method to physics will, I believe, do much more than just improve communication among physicists and between physicists and other scientists. It will accomplish something of far greater importance: it will make it easier to create new concepts, to formulate fresh assumptions" (Carnap 1966, 291). If the argument in this paper is sound, then it suggests that the Carnapian project of the rational reconstruction of a scientific theory like QM is still alive, despite an apparent consensus to the contrary. A viable rational reconstruction might, as Carnap hoped, lead to the creation of new concepts and new rules for QM. Whether this is something that can indeed be accomplished by rational reconstruction is a question worth investigating in further research.

References

Beaney, Michael. 2013. Analytic Philosophy and History of Philosophy: The Development of the Idea of Rational Reconstruction. In *The Historical Turn in Analytic Philosophy*, ed. Erich H. Reck, 231–260. London: Palgrave-Macmillan.

Birkhoff, Garrett, and John Von Neumann. 1936. The Logic of Quantum Mechanics. *Annals of Mathematics* 37: 823–843.

Brandom, Robert B. 2015. *From Empiricism to Expressivism*. Cambridge, MA: Harvard University Press.

Carnap, Rudolf. 1927. Eigentliche und uneigentliche Begriffe. *Symposion* 1: 355–374.
———. 1928. *Der logische Aufbau der Welt.* 2nd ed. Berlin-Schlachtensee: Weltkreis-Verlag. Hamburg: Felix Meiner, 1961.
———. 1934. *Logische Syntax der Sprache.* Engl. translation. London: K. Paul, Trench, Trubner & Co Ltd., 1937.
———. 1939. *Foundations of Logic and Mathematics.* Chicago: The University of Chicago Press.
———. 1943. *Formalization of Logic.* Cambridge, MA: Harvard University Press.
———. 1966. *Philosophical Foundations of Physics: An Introduction to Philosophy of Science.* New York: Basic Books.
Carus, André W. 2004. Sellars, Carnap, and the Logical Space of Reasons. In *Carnap Brought Home: The View from Jena*, ed. Steve Awodey and Carsten Klein, 317–355. Chicago: Open Court.
Demopoulos, William. 2007. Carnap on the Rational Reconstruction of Scientific Theories. In *The Cambridge Companion to Carnap*, ed. Michael Friedman and Richard Creath, 248–272. Cambridge University Press, Cambridge.
Gratzl, Norbert, and Georg Schiemer. 2016. The Epsilon-Reconstruction of Theories and Scientific Structuralism. *Erkenntnis* 81: 407–432.
Hardy, Lucien. 2001. Quantum Theory from Five Reasonable Axioms. arXiv:quant-ph/0101012.
Healey, Richard. 2017. *The Quantum Revolution in Philosophy.* Oxford: Oxford University Press.
Hilbert, David. 1918. Axiomatisches Denken. *Mathematische Annalen*, 78, 405–415. Engl. translation in Ewald, William Bragg (ed.) *From Kant to Hilbert*, vol. 2, Oxford University Press, 1996, 1105–1115.
Horvat, Sebastian, and Iulian Danut Toader. 2023. Carnap on Quantum Mechanics. Forthcoming. In *Rudolf Carnap Handbuch*, ed. Christian Damböck and Georg Schiemer. J. B. Metzler.
Mackey, George. 1963. *Mathematical Foundations of Quantum Mechanics.* Princeton: Princeton University Press.
Potter, Michael. 2000. *Reason's Nearest Kin: Philosophies of Arithmetic from Kant to Carnap.* Oxford: Oxford University Press.
Putnam, Hillary. 1968. Is Logic Empirical? In *Boston Studies in the Philosophy of Science*, eds. Robert Sonné Cohen, and Marx W. Wartofsky, vol. 5, 216–241, D. Reidel, repr. as "The logic of quantum mechanics" in his Mathematics, Matter and Method. Philosophical Papers, vol. 1, Cambridge University Press, 174–197.
Sellars, Wilfrid. 1953. *Inference and Meaning*, 313–338. LXII: Mind.
Stebbing, Susan L. 1932. The Method of Analysis in Metaphysics. In *Proceedings of the Aristotelian Society*, 65–94. Oxford: Oxford University Press.
Toader, Iulian Danut. 2015. Objectivity and Understanding: A New Reading of Carnap's *Aufbau*. *Synthese* 192: 1543–1557.
Von Neumann, John. 1932. *Mathematical Foundations of Quantum Mechanics.* Princeton: Princeton University Press.
Wallace, David. 2020. On the Plurality of Quantum Theories: Quantum Theory as a Framework, and its Implications for the Quantum Measurement Problem. In *Scientific Realism and the Quantum*, ed. Steven French and Juha Saatsi, 78–102. Oxford: Oxford University Press.

Chapter 3
On Hempel on Hempel

Sebastian Lutz

Abstract Hempel publicly abandoned the Received View on scientific theories in the 1960s in favor of a new view. However, Hempel misrepresents his own works within the Received View in a number of his criticisms, and his new view turns out to be identical to the Received View on correspondence rules, observational terms, theoretical terms, and the demarcation between basic principles of a theory and correspondence rules. Hempel's criticism of the assumption of axiomatization has counterexamples in his own previous work within the Received View, and his criticism of the meaning of theoretical terms in the Received View ignores developments he witnessed and discussed.

Keywords Hempel · Received view · Correspondence rules · Abstract calculus · Axiomatization · Observational terms · Theoretical terms · Analytic · Synthetic · Carnap

3.1 Suppe on Hempel on Hempel

In his reminiscence about the 30-year-long supremacy of the semantic view on theories, Frederick Suppe (2000, S102) treats the reader to the violent death of logical positivism:

> The *Received View on Theories* was the epistemic heart of Logical Positivism. Twelve hundred persons were in the audience the night it died. [...] The Received View had been under sustained attack for a decade and a critical mass of main protagonists had been assembled to fight it out. Carl Hempel [...] was expected to present the Received View's latest revision. Instead he told us why he was abandoning both the Received View *and* reliance on syntactic axiomatizations (Hempel 1974). Suddenly we knew the war had been won, and the Symposium became an energized exploration of where to go now.

S. Lutz (✉)
Department of Philosophy, Uppsala University, Uppsala, Sweden

G. Schiemer (ed.), *The Legacy of the Vienna Circle*, Vienna Circle Institute Yearbook 30, https://doi.org/10.1007/978-3-031-80568-4_3

Never mind that the Received View, which had been mainly developed by Carnap and previously relied upon by Hempel, was not the epistemic heart of logical empiricism, simply because a number of logical empiricists took different views on theories (Mormann 2007).[1] Never mind that, to stay in Suppe's violent metaphor, the 1200 conference participants in 1969 did not witness the death of the Received View, but at best discovered its body: As Suppe himself notes (Hempel 1974, 244, editor's note), Hempel's presentation was based on a lecture given in 1966, which in turn was based on a lecture from 1965. What I find most dispiriting is that Suppe presents the philosophical debate not as a successive improvement—if you will even a war—of ideas, but rather a sort of political maneuvering, in which allegiances count more than arguments.

A philosopher of science (as opposed to a historian of philosophy) should care about Hempel abandoning the Received View in the same way that a philosopher of religion should care about Antony Flew abandoning atheism (cf. Oppenheimer 2007): If his criticisms of the Received View are good, they stand on their own. If his criticisms are bad, Hempel's personal stance does not improve them.

As it turns out, Hempel's criticisms are bad, and bad in a particularly puzzling way for an erstwhile proponent of the Received View: Like many critics before and after him (Lutz 2012b), Hempel misrepresents the Received View, sometimes blatantly so, to then criticize the straw man. That Hempel misrepresents his own view is a bold thesis, so I hope that the gentle reader will indulge me while I lay out the evidence, letting Hempel's psychology be the topic for someone more versed in the matter (although I will engage in some groundless speculations in the epilogue).

3.2 Two Views on Scientific Theories

3.2.1 Abandoning the Received View

In 1965, Hempel gave the lecture "The Legacy of Logical Positivism for the Philosophy of Science" as one of the Johns Hopkins Seminars in the Philosophy of Science.[2] His conclusion, in the article based on the lecture, is positive (Hempel 1969a, §6):

> A fair-minded appraisal of the accomplishments of logical positivism should not focus on the bold and naturally oversimplified devices its adherents wrote upon their banners, but on the quality of the detailed logical and methodological studies carried out under those banners[.] Thus judged, logical positivism will be found, I think, to have been a strong and fruitful influence in recent systematic philosophy.

---

[1] Unfortunately, some people who really should have known better have parroted Suppe's assessment (see Lutz 2012b, 77, for a current example).

[2] See the *Virtual Archive of Logical Empiricism*, call number RC 090-75-03. phaidra.univie.ac.at/o:1720472

That same year, and in the following five year span, Hempel gave four lectures in which he abandoned the Received View. He first gave an Isenberg Memorial Lecture in East Lansing in the Fall of 1965 (Hempel 1969b), which was "develop[ed] further, and modifie[d] in certain respects" in his lecture at a Minnesota conference on correspondence rules in May 1966 (Hempel 1970, 142, n. 1). His lecture at the Urbana conference on the structure of scientific theories, on which Suppe reports, was given in March 1969 and published as a "summary-abstract" to "avoid excessive repetition" from the Minnesota conference (Hempel 1974, 244, editor's note). Finally, Hempel gave another lecture at the Fourth International Congress for Logic, Methodology, and Philosophy of Science in Bucharest in 1971 (Hempel 1973b).

In all four of the lectures, Hempel lays out that according to the Received View, a theory consists of two sets of sentences, a set C of formulas of an uninterpreted calculus, and a set R of correspondence rules, which provide the interpretation of the calculus by connecting its terms with observation terms (Hempel 1969b, 49–50; 1970, 146–147; 1973b, 368–369; 1974, §II). Hempel calls this "Schema I" (Hempel 1969b, 49), "the 'standard' or the 'received' philosophical construal of scientific theories", "the standard conception", or "the standard construal" (Hempel 1970, 146; 1974, §II), and the "standard construal, or the standard analysis" (Hempel 1973b, 367).

In the East Lansing lecture, Hempel (1969b, 49) notes: "I have myself repeatedly used a construal of this kind in discussing the status of theories and of theoretical entities; but it seems to me now that, while not strictly untenable, it is misleading in several respects." This is echoed in the Minnesota lecture (Hempel 1970, 147): "I have myself relied on the standard construal in several earlier studies, but I have now come to consider it misleading in certain philosophical significant respects[.]" In a footnote, Hempel gives his article "The Theoreticians Dilemma", written in 1956 or 1957 (Hempel 1958, 37, author's note) as the example of his reliance on the Received View. The Urbana lecture could be considered a further sharpening of his stance, since Hempel (1974, §II) now states: "While I have repeatedly relied on the standard construal in earlier writings, I have come to feel increasing doubts about its adequacy." Thus in East Lansing, the Received View seemed to be misleading, while in Minnesota, Hempel had become more certain and considered it misleading in significant respects. In Urbana, the charge had become straightforward inadequacy, although expressed with some uncertainty. In Bucharest, finally, Hempel (1973b, 377) categorically states that "at least one of the major problems to which the standard conception was addressed, the problem of meaning specification for theoretical terms, rests on a mistaken presupposition and thus requires no solution." In Bucharest, Hempel (1974, 367) also does not point out any longer that he once held the Received View. He instead notes that it was developed (as opposed to merely applied) by "various thinkers [...]; among them, F.P. Ramsey, N.R. Campbell, R.B. Braithwaite, Hans Reichenbach, Rudolf Carnap, Herbert Feigl, and Ernest Nagel."

"The Theoreticians Dilemma" had been revised during a sabbatical at the Center for Advanced Studies in the Behavioral Sciences in 1963–164 for republication in

Aspects of Scientific Explanation. In that collection's introduction, Hempel (1965, vii–viii) states:

> While I still regard the central ideas of the reprinted essays as basically sound, I have naturally changed my views on various points of detail. Where it seemed appropriate, such changes have been indicated in footnotes marked "Added in 1964" or in the Postscripts by which I have supplemented three of the articles.

None of the criticisms Hempel presents in his four lectures beginning in 1965 are mentioned in footnotes in the article's republication, which also does not have a postscript. The collection was published a year later, the same year that Hempel gave his positive evaluation of the Received View in "The Legacy of Logical Positivism for the Philosophy of Science". Thus it is probably fair to say that the four lectures that followed provide Hempel's reasons for abandoning the Received View.

3.2.2 A New View on Scientific Theories

After the Received View is given up, what, then, is the right way of reconstructing theories? At the Minnesota conference, Hempel (1970, §1) provides a better view (see also Hempel 1969b, 50; 1974, 246), with a proviso: In the subsequent discussion, he stresses that his "paper was intended principally as a criticism [and] did not put forward a properly developed alternative" (Achinstein et al. 1974, 261). Hempel's new view reconstructs a theory T as a pair $\langle I, B\rangle$ of two sets of sentences, the internal principles I and the bridge principles B. The internal principles contain only terms from the theoretical vocabulary V_T, while the bridge principles connect these theoretical terms to terms from the antecedent vocabulary V_A. V_T and V_A form a partition of the whole vocabulary. Thus Hempel (1970, 143) suggests that

$$T = I, B, \tag{3.1a}$$

or, "alternatively, and with greater intuitive appeal, T may be construed as the set of logical consequences of the sum of the two sets:"

$$T = c(I \cup B). \tag{3.1b}$$

At the East Lansing conference, Hempel (1969b, 50) does without the deductive closure:

$$T = I \cup B \qquad \text{(Schema II)}$$

As Hempel notes, there "are obvious similarities to the [Received View], but there are also considerable differences; this will become clear in the following amplification and comparison of the two construals."

3.3 Correspondence Rules

The first alleged difference that I want to discuss is, to put it bluntly, silly.[3] In the Minnesota lecture, Hempel (1970, 147) claims:

> Whereas the bridge principles in our initial characterization of a theory are conceived as a subset of the class of sentences asserted by the theory, the status of the correspondence rules in the standard construal is less clear. One plausible construal of them would be as terminological rules belonging to the metalanguage of the theory[.] For this reason, no immediate analogue to (3.1b) is available as an alternative schematization of the standard view.

The problem with this view on correspondence rules is that while a full axiomatization determines "for any sentence *S*, whether *S* is asserted by the theory", and thus "the infinite set of sentences that a proposed theory is intended to assert", the Received View does not treat correspondence rules as sentences in the object languages, so that it cannot assume an axiomatization of the full theory (148–149). (Hempel does not consider the possibility of axiomatizing the calculus in the object language and, separately, the correspondence rules in the metalanguage.)

The reason this criticism is silly is that in "The Theoretician's Dilemma", Hempel's example of him relying on the Received View, Hempel (1958, §§4, 6, 8) treats correspondence rules as definitions, conditional definitions, and more general sentences in the object language, and explicitly uses the conjunction of abstract calculus and correspondence rules as the "interpreted theory" T'. In particular, he uses the Ramsey sentence of T' (Ramsey 1931, §4), which, as Hempel (1970, 149, n. 8) himself notes, presupposes "axiomatization of the entire theory", including the correspondence rules. To make the whole criticism even more puzzling, Hempel mentions Ramsey's investigation explicitly as a contrast to the Received View in this respect, even though he also lists the investigation as one of the "characteristic stages in the evolution of" the Received View (146, n. 4).

What is more, in the East Lansing lecture less than a year earlier, Hempel (1969b, 49) uses the union of abstract calculus and correspondence rules as a description of a theory in the Received View:

> The conception of a theory as consisting of an abstract calculus *C* and a set *R* of rules of correspondence will be briefly represented by the following schema, which construes a theory as the sum of two classes of sentences:

$$T = C \cup R \qquad \text{(Schema I)}$$

In the unpublished lecture "On Theoretical Terms and Theoretical Entities", held in front of the NY Philosophy Club on 21 January 1966 and thus four months before the Minnesota conference, Hempel repeats this construction.[4] Now, Hempel (1970, 142) notes that the Minnesota lecture "modifies in certain respects" the East Lansing

---

[3] I received incredulous stares when showing the following passage at the University of Vienna.

[4] Carl Gustav Hempel Papers, University of Pittsburgh, Identifier 31735062224286, digital.library.pitt.edu/islandora/object/pitt%3A31735062224286, page 3.

lecture, but it would be odd if over the course of four months Hempel had come to the conclusion that the correspondence rules in the Received View are given in the metalanguage, especially since he could have just looked up their status in the article he had revised two years earlier for republication. Furthermore, the criticism is not mentioned in the Urbana summary, and in the Bucharest lecture, Hempel (1973b, 373) states:

> The interpretative sentences of a theory have been variously construed as formulated within the language of science or as sentences effecting a semantical interpretation and thus belonging to a suitable metalanguage. My remarks will be limited to the first, more common, construal[.]

Thus it seems that Hempel himself fielded this criticism only for a short period of time.

As far as the philosophy of science is concerned, though, it is just important that the presumption of Hempel's criticism is also clearly false, since he himself assumed that correspondence rules are expressed in the same language as the theoretical sentences of the theory when working within the Received View, and so did Rudolf Carnap in all of his works since at least "Testability and Meaning" (Carnap 1936, 1937). So Hempel's new view on correspondence rules is not a departure from the Received View, but is the Received View.

3.4 Observational Terms

The current discussions of observational terms usually focus on the difficulty of clearly demarcating observational terms (e.g. Maxwell 1962). This is not Hempel's criticism. He instead states in the Minnesota lecture (Hempel 1970, 143–144):

> The antecedently examined phenomena for which a theory is to account have often been conceived as being described, or at least describable, by means of an observational vocabulary, i.e., a set of terms standing for particular individuals or general attributes which, under suitable conditions, are accessible to "direct observation" by human observers.
>
> [But] the antecedent vocabulary of a given theory will often contain terms which were originally introduced in the context of an earlier theory, and which are not observational in a narrow intuitive sense.

In an comparably elaborate example, Hempel (1969b, 53–54; 1970, 144–145) notes that in the case of Bohr's theory of the hydrogen atom entailing the Balmer formula for the wavelengths emitted by hydrogen, the description of the wavelengths is not given in directly observational vocabulary, but rather the antecedent vocabulary of earlier theories. This is not the first time Hempel uses this example. It also occurs in his textbook *Philosophy of Natural Science* (Hempel 1966, §6.2), a "substantial part of [which] was written in 1964, during the last months of a year [Hempel] spent as a Fellow of the Center for Advanced Study in the Behavioral Sciences" (vii), and thus probably right after finishing the revision of "The Theoretician's Dilemma" for republication, and before giving his endorsement of logical positivism's philosophy of science in the Johns Hopkins seminar. So Hempel arguably did not always see the use of antecedent vocabulary from earlier theories as a problem for the Received View.

As Hempel's criticism in the Minnesota lecture presumes, Hempel (1958, 179) states in "The Theoretician's Dilemma" that in "regard to an observational term it is possible, under suitable circumstances, to decide by means of direct observation whether the term does or does not apply to a given situation." However, Hempel (1958, 42–43) cautions that this characterization

> offers no precise criterion by means of which any scientific term may be unequivocally classified as an observation term or as a theoretical one. But no such precise criterion is needed here; the questions to be examined in this essay are independent of precisely where the dividing line between the terms of the observational and the theoretical vocabularies is drawn.

So Hempel did indeed mean to distinguish between directly observational terms and theoretical terms when applying the Received View, but this does not immediately entail that his arguments actually rely on this. The preliminary discussion (§§1–8) makes up the bulk of the article and focuses on the successive liberalization of the correspondence rules. With the exception of the claim that disposition terms cannot be defined in observation terms (§6), the arguments are independent of any particular dividing line. The only assumption is that observation terms are interpreted, while theoretical terms are not. Analyzing the eponymous dilemma, Hempel (1958, 73) makes this independence explicit: He does not distinguish between observational and theoretical terms at all, but rather between a basic vocabulary V_B and a theoretical vocabulary V_T and states that

> we will have to assume, of course, that V_B consists of terms which are antecedently understood. They might be observational terms, in the somewhat vague sense explained earlier; but we need not insist on this.

Thus Hempel did provide a (vague) demarcation of observational terms, but, crucially, did not rely on it in his analysis. There is even a telling terminological slip-up in which Hempel (1958, 84) speaks of "an observational vocabulary V_B", suggesting that he even thought of observational terms simply as interpreted. And in the East Lansing lecture, Hempel (1969b, 54) was also still clearly aware of this, as he writes immediately after the example of the Balmer formula that,

> while the rules of correspondence, *R*, are often said to connect theoretical terms with observational ones, several writers who have made expository use of that schema have clearly envisaged a less stringent conception that is quite compatible with the one just outlined.

Hempel was one of these writers, as just noted, and so was Carnap, who early on allowed for *R* connecting theoretical sentences with an "inferred primary protocol" involving antecedent terms (Carnap 1931, 437–438) and later simply assumed more and less elementary (i.e., observational) terms (Carnap 1939, §24), as Hempel (1951, 68, n. 11) himself also notes. Hempel (1963, VII) also points out that Carnap (1932) "early rejected the idea of a privileged class of 'protocol sentences' conceived as terminal statements in the process of empirical verification".

Thus Hempel's criticism of the Received View on observational terms is based on a straw man as well, and his new, competing view amounts to the one Carnap held, and he himself held when still working within the Received View.

3.5 The Theory as an Abstract Calculus

In Hempel's account, a theory is in the Received View reconstructed as containing a set *C* of uninterpreted sentences, an abstract calculus. But the use of an abstract calculus is wrong, according to Hempel, both because it is abstract, and because it is a calculus.

3.5.1 The Theory as Abstract

Hempel (1970, 153) states in the Minnesota lecture that

> the internal principles of most scientific theories employ not only "new" theoretical concepts but also "old," or pre-theoretical, ones that are characterized in terms of the antecedent vocabulary. For the theoretical scenario is normally described in part by means of terms that have a use, and are understood, prior to, and independently of, the introduction of the theory.

As a criticism of the Received View as described by Hempel, it is a decidedly odd one: Hempel's *new* view distinguishes between antecedent vocabulary and vocabulary introduced by the theory in question. The Received View, according to Hempel, distinguishes between directly observational and non-observational vocabulary. According to Hempel's description of the Received View, then, two theories could share theoretical terms, since theories can share terms that are not directly observational. So the criticism Hempel aims at the Received View here is, if anything, a criticism of his own new view.

Of course, the previous section established that Hempel himself, in "The Theoretician's Dilemma", relies on a distinction between antecedent and (new) theoretical terms, where the antecedent terms may but do not have to be observational. But even if one assumes that the Received View requires a new theoretical vocabulary, Hempel's criticism of a theory as abstract is a non-sequitur as long as *C* is not fixed independently of the requirement that it contain only new vocabulary, for then *C* can simply be taken to fulfill the requirement by definition. Hence Hempel (1970, 153) adds that

> [t]he assumption, in the standard construal, of an axiomatized uninterpreted calculus as a constituent of a theory [...] suggests that the basic principles of a theory—those corresponding to the calculus—are formulated exclusively by means of a "new" theoretical vocabulary[.]

So Hempel conjectures that in the Received View, the abstract calculus describes the "basic principles of a theory". But apart from it being a conjecture, it is also not clearly defined as long as it is not clear how to identify the basic principles of a theory. Hempel might have had in mind the description of a theory as it is found in textbooks and research papers, like Bohr's description of his model of the hydrogen atom.

With these assumptions in place, the Received View indeed entails that all terms used in a theory are uninterpreted. But the actual Received View does not entail this claim. At best, Hempel here goes against his own admonition to critics of the Received View to "not focus on the bold and naturally oversimplified devices its adherents wrote upon their banners, but on the quality of the detailed logical and methodological studies carried out under those banners" (Hempel 1969a, §6): While sometimes, proponents of the Received View have spoken of the terms of the theory as uninterpreted (see Lutz 2012b, 77–78, for a current example), this cannot have been an assumption in the Received View. For one, the correspondence rules of one theory are often entailed by another theory, for instance a theory describing a measurement device; Carnap (1926, 30), for instance, in a monograph discussed by Hempel (1952, §§10–11, n. 62, 64), relies on a theory about the expansion of materials with temperature to derive a correspondence rule for the notion of length. A theory that is to entail a correspondence rule then obviously has to contain interpreted terms, otherwise it could not connect the uninterpreted terms to interpreted ones.

Furthermore, there is in general no reason for a theory to be restricted to non-observational terms; rather, the basic principles of a theory can connect theoretical and observational terms, and correspondence rules can be used for those terms that have a stronger connection to observation terms than can be inferred from the basic principles of the theory alone. All analyses within the Received View that I am aware of allow for this. Prominent examples are Craig's theorem and the Ramsey sentence, which are simply developed for $T = C \cup R$, that is, the whole of the theory including both abstract calculus and correspondence rules. These analyses are central in "The Theoretician's Dilemma", where Hempel (1958, §9) relies only on the conjunction of abstract calculus and correspondence rules as the "interpreted theory" T'.

3.5.2 *The Theory as a Calculus*

Hempel is against the reconstruction of the basic principles of a theory as uninterpreted, but he is also against their reconstruction as a calculus. Or, more precisely, he is against the simple *assumption* that a theory is formalized in a calculus. In the Urbana lecture, Hempel (1974, 247–248) states:

> [T]he fact that certain philosophical issues may be clarified by an appropriately chosen axiomatization of some particular theory surely does not suffice to justify the idea embodied in the standard construal, that the internal principles of any scientific theory be conceived as being axiomatized in some unspecified way.

This is developed more fully in the Minnesota lecture, where Hempel (1970, 152) states that

> whatever philosophical illumination may be obtainable by presenting a theory in axiomatized form will come only from axiomatization of some particular and appropriate kind rather than just any axiomatization or even a formally especially economic and elegant one.

For Hempel, then, whatever advantages a formalization has can come about only by axiomatizing a theory in a specific way, based on philosophical insights that precede the axiomatization.

As a first criticism, Hempel (1970, 148–149) notes that while an arbitrary axiomatization provides "a general criterion determining, for any sentence S, whether S is asserted by the theory", it "does not provide us with a general method of actually finding out whether a given sentence" is so asserted, because first order logic is not decidable. As a criticism of axiomatization it is remarkably poor, given that, first, there is no better replacement and, second, establishing the precise content of scientific theories is so difficult that the lack of decidability registers barely, if at all. Hempel here seems to be suggesting that because the water bucket may spill, we should instead use paper bags or just let the house burn down.

But Hempel (1970, 152) also gives examples for insights that can come only from appropriate axiomatizations. The axioms for a set of sentences, for instance, can to some extent be chosen at will, and thus do not automatically, simply in virtue of being axioms, express the foundational assumptions of the theory. To do that, the axioms have to be chosen based on philosophical analysis. Similarly, an axiomatization does not identify any terms as primitive, and even a chain of definitions can to some extent (when terms are interdefinable) be chosen at will. Thus both primitive terms and the order of definition must be chosen based on philosophical analysis.

However, this is a proof by example or, in other words, a non-sequitur: That the assumption of axiomatization does not entail answers to some questions does not warrant the conclusion that the assumption entails no answers at all. And when relying on the Received View, Hempel himself only required the assumption that the theory is axiomatized in some way or other to gain philosophical insight.

In "The Theoretician's Dilemma", Hempel (1958, §9) discusses the functional replaceability of theoretical terms. His first method replaces the theory T by a set of sentences containing a sentence $O_1 \rightarrow O_2$ for every deducibility relation $O_1, T \vdash O_2$, in which T is used to derived observation statement O_2 from observation statement O_1. By construction, the resulting set does not contain theoretical terms, and Hempel proves that it establishes all and only those deducibility relations that T establishes. Hempel's proof for this claim relies only on the assumption that the theory is axiomatized in first order logic. This entails, for instance, compactness, so that Hempel can assume that all sets of sentences in the premises can be expressed by T and a single observation statement. The assumption of axiomatization also entails the deduction theorem and its converse, so that Hempel can use that $O_1, T \vdash O_2$ if and only if $T \vdash O_1 \rightarrow O_2$.

The second method that Hempel discusses relies on Craig's theorem, which states that at every recursively enumerable theory in first order predicate logic has a primitive recursive axiomatization (Craig 1953). Craig's theorem also requires no further assumption about how, specifically, the theory is axiomatized.

The third method Hempel discusses is Ramsey's elimination method for theoretical terms (Ramsey 1931, §4), which existentially generalizes on all theoretical terms. The resulting Ramsey sentence entails the same observational sentences as the original theory, which can be proved by only assuming that the theory is axiomatized in predicate logic.

These are of course just some of the features that all axiomatizations in predicate logic have in common. Others are monotonicity, truth-functionality, the possibility of defining a Tarski semantics (also used by Hempel 1958, 82), and so on. Thus, while Hempel does not misrepresent himself in his claim that the analysis of the Received View sometimes assumed any axiomatization whatsoever, he could have found counterexamples to his conclusion that such an assumption can provide no philosophical insight in his own writings.

There are two other problems with the criticism: One is that the Received view is a framework in which to reconstruct theories (Lutz 2012b, §5.2). Thus even if the assumptions of the Received View had no philosophical implications, the actual reconstructions could have. Hempel (1970, 150) even states that Reichenbach's axiomatization of general relativity in the Received View was "philosophically stimulating and illuminating". Finally, it is difficult to see how Hempel's new view can escape his criticism of the Received View: In his new view, $T = c(I \cup B)$, which very much suggests that I (and B) are in some sense axiomatized, since otherwise their deductive closure would be ill-defined. And even though the new view is not supposed to be a fully developed alternative, it is still supposed to show the failures of the Received View.

3.6 Meaning

Hempel's main criticism of the Received View arguably concerns the way it assumes that theoretical terms receive meaning. This is the criticism that Hempel (1973b) singles out in his Bucharest lecture, the basic point being that the Received View gets the role of both the correspondence rules and the abstract calculus wrong. In the Minnesota lecture, Hempel (1970, 149) states that in the Received View

> the meanings of theoretical terms are determined in part by the postulates of the calculus, which serve as "implicit definitions" for them; and in part by the correspondence rules, which provide them with empirical content.

About the correspondence rules R, Hempel (1970, 159) conjectures:

> [R's] designation as operational *definitions*, coordinative *definitions*, or *rules* of correspondence conveys the suggestion that that they have the status of metalinguistic principles which render certain sentences true by terminological convention[.] But such a conception of correspondence rules is untenable for several reasons[.]

This analysis of correspondence rules as conventions is repeated in the Bucharest lecture (Hempel 1973b, §5), in which Hempel (1973b, 370) adds a criticism of the abstract calculus *C*:

> [W]hat exactly is being asserted by this doctrine of implicit definition for theoretical terms? The term 'definition' suggests terminological convention or legislation[. But if] the truth of the theoretical postulates were enforced by terminological *fiat* then the entire theory would be made true a priori; it could be known to be true independently of any empirical evidence—and regardless, moreover, of what interpretations the correspondence rules may assign to the empirical terms.

Thus in the Received View, both *R* and *C* are supposed to be true by convention, but they cannot be if the theory as a whole is to have empirical content.

A good hint that Hempel's overall analysis of meaning in the Received View is flawed comes from his own Bucharest lecture. Hempel (1973a, §4) recounts that originally, correspondence rules were meant to be explicit definitions of theoretical terms in observation terms, but Carnap soon found explicit definitions too restrictive for correspondence rules, so he introduced reduction sentences, essentially necessary or sufficient conditions for theoretical terms. Hempel (1970, 161) recounts:

> Carnap pointed out in his theory of reduction sentences that when a term is introduced [...] by means of several reduction sentences, the latter taken jointly normally have empirical implications.

So Carnap himself pointed out that correspondence rules could not be conventions. Carnap (1936, 451) calls the logically strongest observation sentence entailed by the conjunction of two reduction sentences their "representative sentence", which Hempel (1958, 71) recounts in "The Theoretician's Dilemma". Carnap (1952, 71) later suggested taking the material implication with the representative sentence in the antecedent and the conjunction of the reduction sentences in the consequent as the (analytic) meaning postulate for the introduced term. Hempel (1963, 705) himself discusses this "interesting variant of the method of introducing predicates by reduction sentences" in a text whose first version was written in 1954 (Carnap 1963, 961) and thus before "The Theoretician's Dilemma", but he does not mention it in any of the four lectures.

In his response to Hempel's criticism of the analytic-synthetic distinction (Hempel 1963, VII), Carnap (1963, 24.D) spells out a generalization of this procedure for whole theories:[5] A theory's Ramsey sentence (which generalizes the representative sentence) describes the synthetic component of a theory, and a theory's by now so-called Carnap sentence, a material implication with the Ramsey sentence as the antecedent and the theory as the consequent, describes its analytic component.

To my knowledge, Hempel refers to Carnap's general solution for the analytic-synthetic distinction only once, in a symposium on occasion of Carnap's death (Hempel 1973a, 260). Ironically, the symposium was published the same year as the Bucharest lecture, in which Hempel (1973a, §6) concludes that the Received View's

[5]Carnap's first publication of the idea (Carnap 1958, 245, n. 3) refers for details to this response, to be published in a collection described as "in preparation".

analysis of the meaning of theoretical terms mistakenly assumes that their meaning must be specified linguistically. But this assumption, he states,

> calls for the singling out, from among all the sentences asserted by a theory, of certain sentences which interpret the theoretical terms, and which hold true by virtue of linguistic legislation: and this general notion cannot be made good.

In the Minnesota lecture, Hempel (1970, 159) gives as one advantage of his new view that the "concept of bridge principle as invoked in [his new] characterization of theories does not presuppose the analytic-synthetic distinction and treats the bridge principles [as] on a par with its internal principles". But of course, the same holds true for the Received View.

So Hempel has missed a core development of the Received View. In early versions of the Received View, some proponents may have assumed that the analysis of theories has to proceed through a *specific* kind of formalization—the correspondence rules were explicit definitions and could thus be analytic (as long as each of them defined a different theoretical term), and the abstract calculus could indeed be completely abstract, with the meaning given by the explicit definitions. But at the latest with Carnap's distinction between pairs of reduction sentences on the one hand and their synthetic representative sentences and analytic meaning postulates on the other, there was no such assumption anymore. And with the development of the method of the Ramsey and Carnap sentences, the reconstruction of a scientific theory had become completely untethered from the analytic-synthetic distinction: One can just axiomatize the basic principles of a theory as they are found in textbooks and research papers, resulting in the calculus C, which is not necessarily uninterpreted. For all of a theory's terms that can be independently measured, one can axiomatize the respective measurement procedures, resulting in the correspondence rules R. It is only after the reconstruction that the Ramsey sentence of the theory $C \cup R$ provides the theory's synthetic component and the Carnap sentence provides its analytic component, the meaning of the theoretical terms.

In summary, Hempel recounts Carnap's discussion of empirical implications of correspondence rules, in which Carnap presents an account of constructing analytic meaning postulates from conjunctions of reduction sentences, Carnap replies to Hempel's criticism of the analytic-synthetic distinction by generalizing this account to theories in general, and Hempel in his criticism of the Received View still ignores the central point of Carnap's account: The distinction between correspondence rules and abstract calculus is not the same as the distinction between analytic and synthetic components of theories.

3.7 Conclusion

I have previously been publicly puzzled by Hempel misrepresenting Carnap as relying on first order logic for the reconstruction of scientific theories (Lutz 2012b, 84–85, 87), but the preceding discussion suggest that it is just one example of a large pattern of misrepresentation. Hempel's criticisms of the Received View on correspondence rules, observation terms, and theoretical sentences misconstrue the Received View as he himself assumed it, and contrast it with a new, allegedly more correct view which in these respects is identical to the actual Received View. Hempel's criticisms of the use of axiomatizations in the Received View is simply mistaken, as is exemplified by his own use of axiomatizations in the Received View. And his criticisms of the meaning of theoretical terms and the distinction between correspondence rules and basic theoretical principles in the Received View ignore its development after its earliest versions, a development that Hempel witnessed and commented on when still relying on the Received View.

3.8 Epilogue: On Suppe on Hempel on Hempel

> *Unless we ourselves take a hand now, they'll foist a republic on us. If we want things to stay as they are, things will have to change.*
>
> Tancredi Falconeri to Don Fabrizio[6]

This is the evidence promised at the outset. But why would Hempel misrepresent his own and Carnap's position on scientific theories? Here I can but speculate. Friedman (2003) notes how Hempel was very close to Neurath's naturalism during the time of the Vienna Circle, and how he followed Carnap in focusing on explication while in close contact with him. Under the influence of Kuhn, who joined him in Princeton, Hempel swayed back to a naturalistic description of science. One possibility is that Hempel simply displayed the zeal of the convert, and as a result engaged in motivated reasoning against the Received View. The weakness of his arguments from the metalinguistic status of the correspondence rules and the indecidability of first-order axiomatization point in this direction. But his misrepresentations might also be the result of something like a Gestalt-switch in Hempel's thinking: It might be that his renewed proximity to naturalism was something of a paradigm shift in Kuhn's sense, and Hempel had thereafter difficulties appreciating the basic assumptions of the previous paradigm, the Received View.

Another possibility is that Hempel was aware of the political maneuvering, with war as its extension by other means, that was taking place around the Received View. Recall the description of logical positivism's demise by Suppe (2000, S102):

> Carl Hempel […] was expected to present the Received View's latest revision. Instead he told us why he was abandoning both the Received View *and* reliance on syntactic axiomatizations (Hempel 1974).

And now consider again Hempel's criticism of treating a theory as a calculus. Hempel was not abandoning syntactic axiomatizations, but the assumption of some not further specified axiomatization, *syntactic or not*. Suppe, however, wants to

[6] Giuseppe Tomasi di Lampedusa. 1960. *The Leopard*. New York: Pantheon Books. Translated by Archibald Colquhoun

present the competing semantic view, which includes the set-theoretic axiomatizations by Patrick Suppes, as having won the war, and so he relies on an old trick of rhetoric, the misuse of conversational implicature: By needlessly restricting Hempel's position to *syntactic* axiomatizations, Suppe implicates that Hempel still accepts axiomatizations in general, and thus specifically the semantic ones. But Hempel (1970, 150) introduces his criticism of axiomatizations stating that "Suppes has argued that formalizing and axiomatizing scientific concepts and theories is 'a primary method of philosophical analysis'", to then conclude that "whatever philosophical illumination may be obtainable by presenting a theory in axiomatized form will come only from axiomatization of some particular and appropriate kind rather than just any axiomatization". So Hempel explicitly includes Suppes' axiomatizations in his critique, and thus Suppe misleads, although he does not lie. Political maneuvering again.

So here is a particularly speculative speculation.[7] If Hempel was aware of the political maneuvering and could see the end of the Received View as an accepted approach to research in the philosophy of science, it would have made sense to distance himself as much as possible from the Received View while taking over as much of it as possible into an allegedly new view. For his extant analyses to stay as they were, things would have to change.

I am not endorsing any of these speculations, but I think they are more plausible than that Hempel simply forgot what he had been doing a year earlier. Maybe the most plausible option, however, remains that I am misrepresenting Hempel, for otherwise we would have to accept that Hempel misrepresents Hempel.

Acknowledgments The central contents of this article were presented at the conference *The Vienna Circle—History and Legacy*, University of Vienna, Austria, 26 September 2023. I thank the participants for fascinating discussions throughout the whole conference. Some ideas stem from my dissertation (Lutz 2012a).

References

Achinstein, Peter, Sylvain Bromberger, Robert L. Causey, Carl G. Hempel, Hilary Putnam, and Patrick C. Suppes. 1974. Discussion of 'Formulation and Formalization of Scientific Theories.' In Suppe 1974, 255–265.

Carnap, Rudolf. 1926. *Physikalische Begriffsbildung.* Vol. 39. Wissen und Wirken. Einzelschriften zu den Grundfragen des Erkennens und Schaffens. Karlsruhe: Verlag G. Braun. Translated by Edward Dean and Alan Richardson as "Physical Concept Formation." In *Early Writings,* 1:339–427. The Collected Works of Rudolf Carnap. Edited by A. W. Carus, Michael Friedman, Wolfgang Kienzler, Alan Richardson, and Sven Schlotter, with editorial assistance from Steve Awodey, Dirk Schlimm, and Richard Zach. General editor: Richard Creath. Oxford: Oxford University Press, 2019.

———. 1931. "Die physikalische Sprache als Universalsprache der Wissenschaft." Nominal publication date incorrect. Published in 1932, *Erkenntnis* 2 (5/6): 432–465. Translated by Max

[7] Which I owe to Alexander Linsbichler, who also suggested the introductory quote.

Black as *The Unity of Science*. Psyche Miniatures. London: Kegan Paul, Trench, Trubner & Co., 1934.
———. 1932. Über Protokollsätze. *Erkenntnis* 3 (1): 215–228. Translated by Richard Creath and Richard Nollan as "On Protocol Sentences." *Noûs* 21, no. 4 (1987): 457–470.
———. 1936. Testability and Meaning. *Philosophy of Science* 3 (4): 420–468.
———. 1937. Testability and Meaning—Continued. *Philosophy of Science* 4 (1): 2–35.
———. 1939. *Foundations of Logic and Mathematics*. I:139–213. Foundations of the Unity of Science. Toward an International Encyclopedia of Unified Science 3. Chicago and London: University of Chicago Press. 2 vol. ed.
———. 1952. Meaning Postulates. *Philosophical Studies* 3 (5): 65–73. https://doi.org/10.1007/bf02350366.
———. 1958. Beobachtungssprache und theoretische Sprache. *Dialectica* 12 (3–4): 236–248. Translated by Herbert Bohnert as "Observation Language and Theoretical Language." In *Rudolph Carnap, Logical Empiricist: Materials and Perspectives*, edited by Jaakko Hintikka, 75–85. Dordrecht/Boston: D. Reidel, 1975.
———. 1963. Replies and Systematic Expositions. In Schilpp 1963, 859–1016.
Craig, William. 1953. On Axiomatizability Within a System. *The Journal of Symbolic Logic* 18 (1): 30–32.
Friedman, Michael. 2003. Hempel and the Vienna Circle. In *Logical Empiricism in North America*, Minnesota Studies in the Philosophy of Science 18, ed. Gary L. Hardcastle and Alan W. Richardson, 94–114. Minneapolis: University of Minnesota Press.
Hempel, Carl G. 1951. The Concept of Cognitive Significance: A Reconsideration. *Proceedings of the American Academy of Arts and Sciences* 80 (1): 61–77.
———. 1952. *Fundamentals of Concept Formation in Empirical Sciences*, Foundations of the Unity of Science. Toward an International Encyclopedia of Unified Science. Vol. II,7. Chicago/London: University of Chicago Press. 2 vol. ed.
———. 1958. The Theoretician's Dilemma. In *Concepts, Theories, and the Mind-Body Problem*, Minnesota Studies in the Philosophy of Science, ed. Herbert Feigl, Michael Scriven, and Grover Maxwell, vol. 2, 173–226. Minneapolis: University of Minnesota Press.
———. 1963. Implications of Carnap's Work for the Philosophy of Science. In Schilpp 1963, 685–709.
———. 1965. *Aspects of Scientific Explanation and Other Essays in the Philosophy of Science*. New York: The Free Press.
———. 1966. *The Philosophy of Natural Science*. Englewood Cliffs: Prentice-Hall.
———. 1969a. Logical Positivism and the Social Sciences. In *The Legacy of Logical Positivism: Studies in the Philosophy of Science*, ed. Peter Achinstein and Stephen F. Barker, 163–194. Baltimore: The Johns Hopkins Press.
———. 1969b. On the Structure of Scientific Theories. In *The Isenberg Memorial Lecture Series 1965–1966*, ed. Ronald Suter, 11–38. East Lansing: Michigan State University Press. Reprinted as chapter 4 of *The Philosophy of Carl G. Hempel: Studies in Science, Explanation, and Rationality*. Edited by James H. Fetzer. Oxford: Oxford University Press, 2001.
———. 1970. On the 'Standard Conception' of Scientific Theories. In *Analyses of Theories and Methods of Physics and Psychology*, Minnesota Studies in the Philosophy of Science 4, ed. Michael Radner and Stephen Winokur, 142–163. Minneapolis, MN: University of Minnesota Press.
———. 1973a. Rudolf Carnap, logical empiricist. *Synthese* 25 (3–4): 256–268.
———. 1973b. The Meaning of Theoretical Terms: A Critique of the Standard Empiricist Construal. In *Logic, Methodology and Philosophy of Science IV : Proceedings of the Fourth International Congress for Logic, Methodology and Philosophy of Science, Bucharest, 1971*, ed. Patrick Suppes, Leon Henkin, Athanase Joja, and Gr.C. Moisil, 367–378. Amsterdam: North-Holland Publishing Company.
———. 1974. Formulation and Formalization of Scientific Theories—A Summary-Abstract. *In Suppe* 1974: 244–265.

Lutz, Sebastian. 2012a. *Criteria of Empirical Significance: Foundations, Relations, Applications*. PhD dissertation, Utrecht University. http://philsci-archive.pitt.edu/id/eprint/9117.

———. 2012b. On a Straw Man in the Philosophy of Science: A Defense of the Received View. *HOPOS: The Journal of the International Society for the History of Philosophy of Science* 2 (1): 77–120. https://doi.org/10.1086/664460.

Maxwell, Grover. 1962. The Ontological Status of Theoretical Entities. In *Scientific Explanation, Space, and Time*, Minnesota Studies in the Philosophy of Science, ed. Herbert Feigl, Michael Scriven, and Grover Maxwell, vol. 3, 3–27. Minneapolis: University of Minnesota Press.

Mormann, Thomas. 2007. The Structure of Scientific Theories in Logical Empiricism. In *The Cambridge Companion to Logical Empiricism*, ed. Alan Richardson and Thomas Uebel, 136–162. Cambridge: Cambridge University Press.

Oppenheimer, Mark. 2007. The Turning of an Atheist. *The New York Times Magazine* (November 4, 2007): 36–41.

Ramsey, Frank P. 1931. "Theories." Chap. IX.A. In *The Foundations of Mathematics and other Logical Essays*, International Library of Psychology, Philosophy and Scientific Method. Written in Summer 1929, 212–236. London: Kegan Paul, Trench, Trubner, & Co.

Schilpp, Paul Arthur, ed. 1963. *The Philosophy of Rudolf Carnap*, The Library of Living Philosophers 11. Chicago/LaSalle: Open Court Publishing Company.

Suppe, Frederick, ed. 1974. *The Structure of Scientific Theories*. Urbana: University of Illinois Press.

———. 2000. Understanding Scientific Theories: An Assessment of Developments, 1969–1998. Supplement. Proceedings of the 1998 Biennial Meetings of the Philosophy of Science Association. Part II: Symposia Papers, *Philosophy of Science* 67: 102–115.

Chapter 4
Noncognitive Deliberation in Context

Christian Damböck

Abstract Logical empiricism can be characterized by a double message. All contexts in science and life have a noncognitive side, which is not a matter of right or wrong, but of attitude and value-driven decision making. And the task of science and philosophy is to provide strategies for minimizing the noncognitive input and maximizing the aspects based on rationality, facts, and logic. This logical empiricist stance, in turn, is not only based on a life-reformist attitude and a "scientific humanism". It has very timely implications for a worldview that wants to be based on both democracy and science. This very broad picture of logical empiricism and its political agenda is drawn here against the background of an overview of its historical development.

Keywords Logical Empiricism · Noncognitivism · Rudolf Carnap · Deliberative Democracy

4.1 The Demarcation Between the Cognitive and the Noncognitive

A discussion of noncognitivism, or as I prefer to say here, *noncognitive deliberation*, as developed by the Vienna Circle, the logical empiricists, and especially by Rudolf Carnap,[1] must begin with the observation that "value" here is to be

[1] In this paper, I always try to extend the discussion to the whole group of logical empiricists, which includes not only the Vienna Circle (Stadler 2015), but also the Berlin group around Hans Reichenbach (Danneberg et al. 1994), as well as followers in other parts of Europe and the United States. Nevertheless, the key figure of my discussion will be Rudolf Carnap, with some additional coverage of Hans Reichenbach, Otto Neurath, and Moritz Schlick. This paper is an elaboration of

C. Damböck (✉)
Institute Vienna Circle, University of Vienna, Vienna, Austria
e-mail: christian.damboeck@univie.ac.at

G. Schiemer (ed.), *The Legacy of the Vienna Circle*, Vienna Circle Institute Yearbook 30, https://doi.org/10.1007/978-3-031-80568-4_4

understood in the broadest sense, encompassing not only moral and political norms, but every kind of norm that determines the way human beings approach reality. This, in turn, is based on the specifically limited way in which logical empiricism views the realm of what is considered objective, where it is crucial to note that for logical empiricists "objective" is meant in the specific sense of being intersubjectively accessible. *To be objective means to be observable by everyone, from exactly the same source*, even if often mediated and indirect, for example when we refer to events in the past.

There are only two possible sources in logical empiricism for judgments that can be considered objective, because they are always testable in an intersubjective environment of any number of people who share these sources. First, empirical judgments of a "physicalistic" nature or protocol sentences, i.e., judgments that concern observable properties of observable spatiotemporal objects. Second, analytic judgments that can be proved within a given environment of axioms and rules of derivation. But even these statements being objective insofar as their truth entirely and directly depends on something that is externally given in an obvious way and therefore does not depend in any significant way on the attitudes of those who establish the truths are accepted by means of a decision and insofar have a *noncognitive* side. And this means that, according to the lore of logical empiricism, we must distinguish six different domains of noncognitive attitudes that comprise all regions of science and life:

1. *Protocol statements and analytic statements.* To counter a prominent misconception of logical empiricism, it is important to note that for this philosophical movement there are no statements that can be considered absolutely certain. They "put all concepts of the language of science into the 'floating net' [...] no 'solid rock' remains, and there are no unshakable protocol sentences" (TB September 2, 1940). Protocol sentences and analytic statements are the foundations of science, because they directly address exactly what is objectively given. Nevertheless, they are fallible and subject to error, hallucination, and fraud. Thus, it is a matter of attitude and subjective evaluation to decide whether to trust reports of empirical events or formal proofs. These statements *match* the reality, but only if they really do: and whether they *really* do, whether we trust them, is something we have to decide, either on the basis of our own observation, or on the basis of testimony, or on some other form of evidence. No matter how we actually decide, insofar as their validity is based *on a decision*, even the absolutely cognitive statements as statements actually have a noncognitive side.
2. *Scientific theories.* Although they can be seen as more or less empirically justified and tested, scientific theories always leave room for noncognitive choice. There may be empirically equivalent alternatives between which only the scientists' attitudes allow a choice (van Fraassen 1980, ch. 3), and sometimes there

(Damböck 2018, 2022a, b, d, e, 2024a, c). I am indebted to Martin Kusch, Sebastian Lutz, Anne Siegetsleitner, Friedrich Stadler, Alexander Reutlinger, and Marta Sznajder for their comments. This work was supported by the Austrian Science Fund (FWF grant P34887).

may even be noncognitive reasons for preferring a theory despite the fact that it is empirically weaker than the rejected alternative (Kuhn 1996, ch. XIII). Logical empiricism is anti-realist in this respect, because it denies that any scientific theory can ever be considered definitely true, no matter how strongly it is supported by empirical and logical evidence.

3. *Logics, languages, and scientific frameworks.* In contrast to ordinary language philosophy, which tended to view all kinds of frameworks as parts of nature that humans could not or should not intentionally change, the logical empiricists were clearly committed to the idea that both logics and (everyday and scientific) languages, as well as all kinds of scientific frameworks are basically human inventions that humans can freely change again. Nothing can be considered fixed here, although different frameworks can always be shown to serve certain purposes better or worse than others. Frameworks may be more or less useful or fruitful in one context or another. This usefulness of a framework is not something that can be decided absolutely and once and for all, because the decision inevitably depends at least partly on the feelings and attitudes of those who decide. (Carnap 1950)
4. *Moral and aesthetics, legal systems, and politics.* Values in the classical sense, which determine our moral, aesthetic, and political preferences as well as the legal systems we adopt, are a matter of noncognitive choice, although they depend on cognitive update, which singles out the causal consequences of a valuative choice and evaluates the consistency of a value system. Cognitive information about such empirical and logical connections can and should influence our value choices, but they are always ultimately dependent on our attitudes. (Carnap 1934; Damböck 2022e)
5. *Art and Literature.* All fictional products of the human mind are obviously noncognitive in nature, although they may use cognitive knowledge as a guiding principle in one way or another. Logical empiricism, to be more precise, does not necessarily defend particularly irrational or expressionist varieties of art and literature. Rather, there are affinities between some representatives of logical empiricism and the movements of New Objectivity (Neue Sachlichkeit), and the Bauhaus, with their strong tendencies toward abstraction and structuralization, on the one hand, and an objectifying approach to art, on the other, in which impartial visualization through photography and collage is as highly valued as scientific and technical literature, which provides important stylistic elements for fiction. (Dahms 2004, 2021; Damböck 2024b)
6. *Human life, decision making, and action.* Life is neither identical with science nor with politics. It is full of aimless games (Schlick 2006, 43–50), mere pleasure or suffering, and it is framed by a chain of decisions of the experiencing individual that typically have no serious consequences outside the aimless game of life. These decisions are highly indeterminate by default, and involve less a clear, laboratory-like scenario of ranked options and comprehensively described causal chains and consistency tests than "auxiliary motives" (Neurath 1981, 57–68) and the will to improvise. The illogical, the "immoral" and crooked sides are what make life enjoyable for many and most. Logical

> Empiricism utterly rejects any attempt to subordinate individual life to any overriding goals or duties. It defines individual happiness as an end in itself, and thus abandons all religious, philosophical, or political attempts to shift the meaning of life to something outside the individual's immediate experience. (Schlick 2006; Carnap 1932; Neurath 1981, 103–129; Stuchlik 2011; Linsbichler and da Cunha 2023) In this sense, logical empiricism maximizes the noncognitive status of life and its meaning. The other side of the coin is that the scientific world conception is primarily intended as a fundamental life reform, placing science in the center of life as a universal aid. (Carnap 1998 [1928], XIII–XVI; Verein Ernst Mach 2012 [1929], 89–90) We must not trust fairytales, conspiracy theories, and hoaxes, but we must trust science, facts, serious predictions, careful explications, and logical evaluations. This will not prevent us from celebrating aimless games and enjoying the illogical and "immoral" side of life, but whenever it really matters, we can stand firmly on the shoulders of science: "*The scientific world-conception serves life, and life embraces it*" (Verein Ernst Mach 2012 [1929], 90).

All products and activities of the human mind are therefore loaded with values that the logical empiricists regard as noncognitive, but the overall strategy of logical empiricism has never been to blow up these noncognitive aspects in the vision of a world in which everything is permeated by subjectivity and various kinds of arbitrariness. It was rather the opposite. The 'scientific world conception' seeks to position science and the cognitive as a model that should penetrate as many areas of life as possible, while pushing back irrationalism, mysticism, and theology. But the logical empiricists also recognized that the real enemies of science and rationality are those who blur the boundaries between the cognitive and the noncognitive and present attitudes in cognitive/theoretical disguise (Carnap 1932, section 7).

Let us now, for the sake of historical reconstruction, turn the tables and leave only the existence of what the logical empiricists call the cognitive as an unquestioned consensus—no reasonable scientifically minded person will deny the objective merits of simple empirical observations and formal proofs—and put in question the entire remaining realm of values or value-laden judgements. A "value" here is simply anything that is considered non-cognitive by the logical empiricists. These values, and the respective judgments that affirm or deny them, have in the past often been considered cognitive, or at least objectively justifiable in one way or another. So, to find out how they came to be considered noncognitive and (which is not the same thing, as we will see) non-justifiable, we will have to span a very wide historical arc, starting with those views that once took all values to be strictly cognitive and objectively determinable.

4.2 Value Objectivism and Beyond (from Hume and Kant to Neo-Kantianism)

The historical pioneers of noncognitivism and noncognitive deliberation fought from the beginning against a classical metaphysical worldview that they attributed to philosophers who had lived before and who were now being overtaken by their new ideas. This metaphysical worldview that supposedly dominated human thought from antiquity to the Enlightenment may be partly or wholly a myth and it may turn out that philosophers have been challenging this compact picture in one way or another all along. Nevertheless, there is a strong conception of value-objectivism that has been defended in whole or in part by philosophers until today, and that can therefore serve as a foil for the discussion here.

Value Objectivism In the value objectivist's (Platonic) heaven, everything that the logical empiricists consider noncognitive is fixed and settled once and for all. The easiest way to imagine this, of course, is a strong metaphysical worldview that takes all values to be given by means of eternally fixed Platonic ideas that can be reconstructed by science and/or philosophy by means of some concepts that reflect the Platonic ideas, typically in an axiomatic setting. Leibniz's *Universal Characteristics* or Spinoza's *Ethics* might be seen as paradigmatic examples for this world view. It should also be noted, however, that in order to become value-*objectivist*, any such construction must be combined with the idea that the respective fundamental axioms or stipulations, which then allow for the derivation or calculation of all values/concepts, are eternally fixed and unchangeable. As soon as we abandon this basic assumption, any such rationalist construal immediately becomes metaphysically innocent, although it then of course has to face the challenge of providing sufficient motivation for its usefulness.

Hume's Noncognitivism If noncognitivism is understood as the purely analytical view that value statements are not truth-apt, because they do not correspond to an external reality that makes them true or false, but merely reflect the emotions or attitudes of those who adopt them, then the actual founder of this view is David Hume, who formulated this position in the *Treatise* (Book III, Part I). However, Hume's conception also shows that the mere understanding of value statements as not being truth-apt not necessarily leads to a rejection of value objectivism. Hume sees the production of values as the result of very specific emotions, namely our "moral sense", which can be seen as a fairly stable thing. If the moral sense is something that depends only on our status as human beings, and therefore must be due to every human being in exactly the same way, then this Enlightenment view of values is hardly any less of an objective everyone-sees-the-same-thing approach than the rationalist proposals of Leibniz or Spinoza. But there is another reason why Hume's approach must be seen as quite different from the logical empiricist conception of value, because Hume limits his noncognitivism to the realm of moral sentiments, i.e., to item (4) of the list from the previous section, while all the rest is obviously taken by Hume to be objectively determined. In other words, Hume's worldview is strongly deterministic wherever logical empiricism reveals indeterminism and room for fallibilism and subjective attitude.

Kant's Transcendental Philosophy Hume opened the door to a noncognitivist view only in the limited field of moral and aesthetic values, while keeping the rest of the epistemic realm strictly objective. This was only corrected by Immanuel Kant, who totalized Hume's shift to the noncognitive in his so-called Copernican turn: "Up to now it has been assumed that all our cognition must conform to the objects […]. Hence let us once try whether we do not get farther […] by assuming that the objects must conform to our cognition" (KdrV, B XVI). For Kant, all theoretical concepts were no longer something that we receive from somewhere outside, but we, the human mind, construct them and only then bring them into contact with reality, and this means, for Kant as for the logical empiricists, either the observable spatiotemporal reality or the reality of formal proof. But Kant, like Hume, also trusted the Enlightenment idea that the human mind is an absolutely stable, historically unchanging thing that also corresponds in some way to reality. This led him to believe that there is a capacity for "transcendental" reasoning, which ultimately implies that all but purely empirical or purely logical concepts and claims are to be justified a priori, by means of what he called "synthetic judgments a priori". In other words, what is noncognitive for the logical empiricists is synthetically a priori for Kant. His potentially complete noncognitivism falls back into a complete cognitivism.

Brentano's Noncognitivism An interesting variation on the objectivist views just outlined is Franz Brentano's value philosophy, which elaborates on Hume's view that value statements are not truth-apt. While sharing this view, Brentano and his followers do not simply ascribe the production of value statements to a "moral sense" that everyone must possess, so that everyone knows what is moral or immoral by means of this stable capacity of evaluation. Rather, Brentano believes that only some people have the ability to produce "correct" and reasonable emotions, which then lead to correct value judgements, whereas the rest of humanity is simply "value blind" in that it lacks the specific emotional abilities needed to correctly evaluate morally sensible situations. (Kraus 1937, 439–441; Damböck 2022e, 501) Despite its absolutism, this position offers an important feature that is also relevant for logical empiricism, namely the idea that our value-producing attitudes may be clouded by emotions that make us fail to see how we would really evaluate a moral problem after we have cooled down and managed to look at the issue with a clear mind. To emphasize this difference, Carnap in his later work preferred not to use the term "emotivism" for his philosophy of values, emphasizing the importance of finding our attitudes that support a value statement on a long term basis and are not clouded by "momentary emotions". (Carnap 1963b, 1000, 1009; Damböck 2022e, 508–510) Thus, for Carnap, emotions must be stable, and this is close to Brentano's correct emotions, although Carnap does not, of course, think that these stable emotions necessarily produce the same value statements for everyone, because they only reflect the property of being unclouded by transient feelings that we might later regret. In any case, for Carnap, people who evaluate a situation equally carefully can still arrive at different value statements. (Carnap 1963b, 1008)

Neo-Kantianism The second half of the nineteenth century brought a wave of relativism to science and philosophy. (Damböck 2017, 22–30) In the German-speaking world, the most significant outcome of this was historicism, here in the sense of the idea that what was once considered in the Enlightenment to be stable, transcendental, and historically unchanging human nature is in fact highly context-dependent and also changes in history. Thus the transcendental, the moral sense, even reason, can change as historical circumstances change. (Daston 2001) This great wave of relativization also led to varieties of historically relativized transcendental philosophy, such as can be found in the Marburg and Southwest German schools of neo-Kantianism. These neo-Kantians, however, not only often held views in which reason as such and moral values were still considered stable and historically unchanging; when they did indeed historically relativize aspects of the transcendental, they did so only relative to a particular existing cultural context. In other words, this relativism left no room for subjective differences in evaluating a context, and there could be only one correct set of values representing the specific context X. (Windelband 1921, II, 59–135) With respect to moral issues, this led to a peculiar form of relativism in which values could change over time, but still had to be fixed for any particular historical situation. Georg Simmel, who defended such a view, accepted that each person has their own moral sense, which may evaluate a situation differently from the rest of the group, but this evaluation is true only if it converges with the group. Therefore, in 1913, even an "antimilitarist is morally committed to armed service, although his subjective moral consciousness rejects it[, …] because individuality is co-determined by it or includes it as a not at all eliminable moment that this person is for instance citizen of a certain state" (Simmel 1913, 151, my translation).

4.3 Noncognitivism and the German Youth Movement

The neo-Kantian understanding of values relativized them, but only in relation to contexts, not to individuals. Therefore, in a given historical context or epoch, the values that represented that context were still fixed, the individual had no choice. An antimilitarist in 1913 had to face the fact that their attitude was simply wrong because it did not reflect what was at stake in their historical realm. A person with a proper moral sense had to be able to figure this out and recognize what the correct values of their context actually were in order to act accordingly. This moral sense had to be an ability to understand a historical epoch and to establish its moral characteristics in the form of a canon. Other less sensitive representatives of the epoch could simply trust the experts and share the canon. But this led to an obvious challenge for the contextualizing approach to values, for it became questionable how epochs could ever change. What is it that gives rise to a new epoch? What kind of moral sense can represent and reflect such a change? It would be rather pointless to believe that epochs change tacitly and people happen to notice that new values have emerged, in the sense of a revelation of something previously hidden, because it is the people themselves who represent the values. The epoch *is*, in a sense, nothing but a set of values and habits. So what if the antimilitarist of 1913 is just a pioneer of a new epoch who already realizes something that the rest of the group cannot yet see?

One aspect of the German life reform and youth movement of the early twentieth century was based on exactly this possibility. People such as the reform pedagogue Gustav Wyneken established the youth as society's capacity to constitute a new epoch. While the adult's duty is to defend the given canon as the cultural property of their group, the youth's duty is the opposite: to question the inherited values, to invent new ones, and thus to establish new canons and epochs. (Damböck 2022b, 166–168) But not everyone in the youth movement shared this view, which some found far too insensitive to the will of the individual. If the youth have only the duty to establish a new canon, then the individual will is rather irrelevant, unless it contributes pieces to the newly growing collective will. What such member of the German youth movement as Rudolf Carnap and Hans Reichenbach had in mind, however, was an entirely new way of living that encompassed not only the youth but the whole of society and entirely discarded the idea of an epoch or collective will, replacing it with an understanding of moral that was exclusively based on the moral attitude of the individual.

> The supreme moral ideal is exemplified in the person who determines his own values freely and independently of others and who, as a member of society, demands this autonomy for all members and of all members. (Reichenbach 1978 [1913], 109, Reichenbach's emphasis)

Morality was moved to a meta-level here. The only demand formulated was to be free to choose one's own values—and to demand the same from everyone else. This freedom came at a price, namely abstaining from alcohol and other intoxicating substances, and adopting a lifestyle that made one sensitive to contemporary events. Other than that, however, there was absolutely no tendency in this metaethical doctrine, for example, to try to persuade others to adopt one's values, or to reach any kind of consensus or compromise. If others simply act as they wish and give me the same freedom, then the world is in an ideal state of moral perfection.

It is striking that this moral principle was formulated just one year before the outbreak of World War I and only 25 years before the Holocaust. How, one is tempted to ask, are we to deal with this understanding of morality when my peers decide to act in a way that is harmful to me and others in the group? Of course, one could try to fix the general principle by introducing some kind of non-aggression pact: act freely, but do not harm others. But this seems rather arbitrary, given that we are not asking others to interact at all, only to let their egos float freely. And this discrepancy became clear to the members of the youth movement as their bitter lesson from the world's descent into chaos and destruction. (Werner 2022; Damböck 2022b, 171–174; 2022c, 43–57) The initial idea, of course, was that morality is not something that can be seen as fixed from the outside in any way, but depends only on the individual's own attitudes and feelings. But this idea in itself is not all we need to build a moral philosophy that is concerned with the noncognitive status of values. There are two other important questions that need to be asked. How do we first identify our own attitudes? And how do we deal with value conflict?

How to Identify Our Own Attitudes If values are noncognitive attitudes, then not only must we reject the idea that they are imparted to the individual from some external source, but there is no reason to think that the individual owes these values internally in any determinate and unambiguous way. Consider the following scenarios. (1) When we are emotionally aroused by an event, we tend to develop very different moral attitudes than we would from a distance, after the transient emotions have cooled and we see the whole scenario in a more impartial way. Should we follow the emotionally charged judgment or the more impartial one? (2) Moral and political decisions always tend to have more or less far-reaching consequences. There may be a fundamental difference between a decision made on the basis of full knowledge of the expected consequences of the various options for action and a decision made without any knowledge of these consequences, or even with misconceptions about them. (3) Different moral principles adopted by the same individual may be contradictory, and there may be a lack of simple general principles from which the more specific ones can be derived. Should we adopt moral principles even if they are contradictory or lack general guiding rules? (4) If we have a number of moral principles, even if they are consistent and follow from some simple basic norms, it may be difficult to weight and schedule them. Which principle do we prioritize in a situation? What weight do we give to different principles? How much are we willing to pay to achieve different goals?

How to Deal with Moral Conflict Our moral worldviews, no matter how we deal with the internal issues mentioned above, depend heavily on how our fellow human beings deal with the issues at hand. The plans of others may be perfectly compatible with ours, even if they are quite different. This may have been what Reichenbach and his allies in the youth movement had in mind. We decide to dress differently, listen to different music, read different books, adopt different sexual preferences, speak different slang, and travel to different regions of the world. And none of this might affect any of our fellow human beings in any negative way. But we are not yet talking about issues of real moral or political value. What if our fellow human beings decide to favor a political or economic strategy that we totally reject, because we think it might have serious negative effects or prevent important positive effects from occurring? What if our fellow human beings adopts measures that directly conflict with our own plans?

Noncognitivism makes questions like these relevant in the first place, because if we outsource moral questions to an external authority, then the answer depends entirely on them, and may have nothing to do with emotional stability, or causality, or logical consistency, or weighting and scheduling, or even moral conflict. Faced with noncognitivism, we are bound to take a stand here, if only to the extent that this stand means that we *ignore* all such questions. Reichenbach's initial strategy seems to have been just that. He recommended that all internal questions and moral conflicts be ignored, advising people to go with their gut and let everyone else do the same. The failure of this strategy became apparent as soon as intractable conflicts arose in society and people realized that morality and politics involved more than just innocuous questions of lifestyle that could be resolved in an environment of happy agreeing to disagree. There emerged two different strategies to make noncognitivism fit for a reality of conflict and complexity, a fascist and a democratic one.

The fascist view, adopted by proponents of a "revolution from the right" such as Hans Freyer and Carl Schmitt, recommended that in an environment of conflict, one should simply try to fight the enemy and make one's fellow group members share one's values by any means available. And this meant that conflicts were not to be resolved, but carried out with the result of the survival of the fitter. On the level of internal moral issues, it meant that in order to get others to share one's values, one should use propaganda. The goal was to get others to overlook their own feelings, as well as factors of fact, causality, and consistency, whenever these factors might prevent others from sharing one's value system. In other words, the key concept of fascism was *irrationalism*, because to function properly, a fascist system must bring the entire population into line through propaganda that causes people to abandon rational decision-making. (Damböck 2022b, 175–177)

The democratic strategy, not surprising, took the opposite path. For democracy to work, two conditions must be met. At the level of society, there must be what Reichenbach himself later characterized as the "democratic principle" he now adopted in contrast to his earlier "anarchist principle": "*everybody is entitled to set up his own moral imperatives and to demand that everyone follow these imperatives*" (Reichenbach 1951, 295, Reichenbach's emphasis). This is meant here in the sense of a deliberative rule that asks the individual to be equally willing to "learn from the group" and "to steer the group in the direction of his own volitions". (Reichenbach 1951, 300) At the level of internal questions, the democratic strategy involved maximizing rationality, along the lines of the four cases mentioned above: trust only long-term and stable emotions, pay attention to all available information on causal consequences of a possible action, try to develop a consistent value system where every single norm follows from a small set of fundamental norms, and strive for a consistent strategy that weights and schedules your principles.

4.4 Noncognitive Deliberation: Systematic Aspects

Noncognitivism with respect to moral statements begins with the *formal* claim that the acceptance or rejection of all value statements ultimately depends on the attitudes of the individual. Once we reject foundationalism, this becomes a truism, for absolutely every statement is ultimately noncognitive in that its acceptance or rejection depends on an attitude in the way illustrated in Sect. 4.1. Thus, noncognitivism with respect to moral statements becomes a meaningful theory only when we add concrete *material* conditions and criteria that our attitudes must satisfy. Here, unfortunately, the logical empiricists have often been sloppy or have provided only a limited catalogue of criteria. Oddly enough, the reception of noncognitivism to this day is often based exclusively on two texts (Ayer 1946, 102–120; Carnap 1935, 22–26), with Ayer's account probably being a direct elaboration on Carnap's all too sketchy view, which does not even mention the material aspects of non-cognitive deliberation as outlined in this chapter. Gross misunderstandings within and outside of the camp of logical empiricism were predestined with these texts.

But putting together the most relevant writings that include (Carnap 1934, 1937, 1944, 1950, 1963a, 81–84; 1963b, 999–1013; 2017; Carnap and Hochkeppel 1967; Reichenbach 1951, 276–302; Kelsen 1929, 1932, 1937, 1955; Neurath 1996, 248–261) the following picture can be drawn. Note again that all of these principles are fundamental norms that do not follow automatically from the formal presupposition of noncognitivism, but are something that we add on a material level, i.e., noncognitivism would be compatible with the absence of any of these norms, and only by adopting them do we obtain what logical empiricists have called "the scientific world-conception" and "scientific humanism," respectively.

1. *Long term emotions*. What our attitudes actually are, and therefore what values we can actually adopt, must first be excavated within ourselves. We must not trust transient emotions, but only "satisfaction in the long run" (Carnap 1963b, 1009). This is a kind of transcendental approach, because what we need to find out are those intuitions that are really at the back of our minds, and which must not be clouded by emotional outbursts. The difference with Kant is that these pure intuitions can still differ from person to person (Carnap 1963b, 1008).
2. *Causal consequences of a decision, side effects*. Whenever a decision is to be made, we should gather as much information as possible about the causal consequences of different courses of action. Our decisions and values should be made only in the context of a thorough evaluation of these causal consequences. Especially important are considerations of collateral damage. A decision may be able to achieve a particular goal, but the side effects may conflict with other values and goals.
3. *Consistency and derivability from fundamental norms, explication*. Values must not be in contradiction to each other and we also should try to find a small set of fundamental norms from which all specific value statements can be derived. In addition to this, value statements may also become subject to further explication that introduces more subtle criteria for "value functions", along the lines of (Carnap 2017; cf. Damböck 2024c).
4. *Side effects, scheduling and weighting*. Actual decisions involve careful strategies of prioritization, scheduling, and weighting. How do we rank priorities? What level of side effects is acceptable? What budget are we willing to spend on what goal?
5. *Moral disagreement I: deliberation*. Even if two people have maximally explored all rational aspects of values in the sense of (1)–(4), it may turn out that they still disagree on moral and political issues. Therefore, deliberation also involves

strategies of education and persuasion, as well as those aspects of propaganda that paint colorful pictures of the defended opinion without distorting or hiding any facts. Roundtable discussions are needed in which people can argue along the lines of (1) through (4), but also give their own subjective side of their value judgments—e.g., through biographical anecdotes, parables, or historical knowledge—and propagate the desirability of their preferred goals. The discussion may become heated. At the end of the debate, disagreement may remain. Then it is a matter of deciding whether the disagreement is something that, though undesirable, will not hinder further coexistence, or whether it is so fundamental in nature, e.g., because the opponent wants to seriously harm or even kill me and my group, that the only option left is to fight for my values.

6. *Moral disagreement II: compromise*. While deliberation is an important aspect of democracy, it is not always necessary or possible to consensually deliberate at a level of roundtable discourse. As much as democracy depends on deliberation, it also deserves the possibility of peaceful coexistence and parallelism of cultures, and a certain amount of "agonism" is not necessarily a threat to democracy but may even be a productive cultural force (Mouffe 1993). At the other end of the democratic spectrum is the need to simply prevent certain actions that may harm others in the group. Thus, democratic institutions include elections and the "majority vote principle", which implies equal freedom for as many individuals as possible:

> The idea underlying the principle of majority is that the social order shall be in concordance with as many subjects as possible and in discordance with as few as possible. […]
>
> The principle of majority, the greatest possible approximation to the idea of freedom in political reality, presupposes as an essential condition the principle of equality. For the view that the degree of freedom in society is proportionate to the number of free individuals implies that all individuals are of equal political value and that everybody has the same claim to freedom, that is, the same claim that the collective will be in concordance with his individual will. […] as many as possible shall be free, […] the mere number of free individuals is decisive. This synthesis of freedom and equality is at the basis of the democratic idea concerning the relationship between the social order (as the collective will) and the individual will […]. (Kelsen 1955, 25)

Deliberation is certainly a good thing, but it is primarily a matter of reaching a consensus on matters of rationality, fact, and science. After agreement is reached on all these matters, there may still be wide disagreement on the level of political and moral attitudes. Here, the "principle of majority" dictates the development of a compromise that prevents as few individuals as possible from acting as they wish.

4.5 Noncognitivism Versus Objectivism in the Twentieth Century

To consider the place of the political and moral philosophy of logical empiricism in twentieth-century thought, it is important to limit the political and moral philosophy of logical empiricism to its theoretical foundations as described in the previous section. This means excluding (a) all more specific political commitments, such as Austromarxism or social democracy, and (b) the question of political commitment in the public sphere. What is often called the "left wing" of the Vienna Circle adopted a more explicitly social democratic and publicly engaged stance (Uebel 2004, 2020; Yap 2024), whose explicitly political nature was not shared by the rest of the group. Nevertheless, there is a broad consensus at a more fundamental level of politically engaged philosophy of science (Romizi 2009), which includes a broad acceptance of the general views outlined in the previous section. I will focus here only on how these theoretical foundations were received by the logical empiricists, the historians of the movement, and by representatives of competing philosophical movement. For this purpose, it is useful to distinguish between *four generations* of philosophers.[2]

First Generation (Born Before 1900) These philosophers, who all experienced World War I, some of them as soldiers, and who published some of their most important writings in the 1920s and 1930s, include the pioneers of the movement—Schlick, Carnap, Reichenbach, Neurath, Kelsen, Friedrich Waismann—as well as important role models such as Ernst Mach and Max Weber, and critics such as Ludwig Wittgenstein, Susan Stebbing, Max Horkheimer, John Dewey, and C.I. Lewis.

Second Generation (Born Between 1900 and the Middle of the 1920s) Most of them experienced the Second World War, some of them as soldiers; includes students of representatives of the first generation, among the followers of logical empiricism—Carl Gustav Hempel, Herbert Feigl, Alfred Tarski, Kurt Gödel, Rose Rand, Charles L. Stevenson, Wolfgang Stegmüller—as well as its critics—W.V.O. Quine, Ernest Nagel, Nelson Goodman, Thomas Kuhn, Alfred J. Ayer, Frank Ramsey, Richard M. Hare.

Third Generation (Born Between the Middle of the 1920s and the End of the 1950s) The postwar generation that includes, among nonstandard followers and critics of logical empiricism such as van Frassen, Hilary Putnam, Ian Hacking, or Nancy Cartwright, the first generation of historians of logical empiricism, in Europe Andreas Kamlah, Hans-Joachim Dahms, Rainer Hegselmann, Michael Heidelberger, Friedrich Stadler, Thomas Uebel, and Thomas Mormann, in the United States

[2] For the historical background, see the writings cited elsewhere in this paper and (Siegetsleitner 2010, 2014; Reisch 2005, 47–53, 382–384; Richardson 2007; Carus 2021, 2022a, b; Dewulf 2022).

Michael Friedman, Nicholas Rescher, Wesley Salmon, Richard Creath, André Carus, J. Aberto Coffa. Representatives of competing movements include especially the Frankfurt school (Wilhelm Habermas).

Fourth Generation (Born After 1960) Scholars who are currently active in academia.

Given this list of generations and some representatives of them, the following three theses can be formulated:

1. *Logical empiricists of generation 1, 2, and 4 typically share the logical empiricist view on noncognitive deliberation.* It is striking that despite the wide range of disagreements at the level of theoretical philosophy and the very different political views that some logical empiricists held, they all agreed broadly on the theoretical foundations of noncognitive deliberation. This is true for the representatives of the first and second generations, including Schlick, Carnap, Neurath, Reichenbach, Hempel, Feigl, Rand, and Quine; and it is true today for most of the scholars of logical empiricism, i.e., the fourth generation.
2. *Third generation logical empiricists and historians of the movement leave the boat.* Two groups must be distinguished here. On the one hand, students of first and second generation logical empiricists such as Putnam develop an approach to philosophy that widely rejects key ideas of logical empiricism and not only adopts very strong metaphysical varieties of scientific realism, but also quite polemically rejects noncognitivism and recommends replacing it with value absolutism of some form. (Putnam 1983, 2002; cf. Linsbichler in preparation) The second group includes scholars from the U.S. such as Friedman and Carus, but otherwise mainly European scholars, such as Kamlah, Dahms, Hegselmann, Heidelberger, Mormann, and Uebel. Again, there are two quite different views to be noted here. First, those scholars who have spent their career mainly in Germany (Kamlah, Dahms, Hegselmann, Heidelberger, Mormann) simply reject noncognitivism out of hand, objecting that it leads to a devastating form of value relativism that renders all value statements entirely arbitrary and thus implies that logical empiricism cannot meaningfully argue against Nazis and other enemies of democracy and the free world (Kamlah 1977; Hegselmann 1979, 59–66; Dahms 1994, 115–120; Heidelberger 2024, section 4; Mormann 2006, 2007, 2010, 2016). Second, scholars from the English-speaking world (Friedman, Carus, Uebel), whose interpretation is quite different in that it defends aspects of noncognitivism, but also adds that in order to achieve a proper—non-relativist—form of rational decision-making, one needs to add an understanding of "practical rationality" or "communicative rationality" that is close to the Frankfurt School in general and Jürgen Habermas in particular; such an understanding can either be found directly in the writings of the first generation of logical empiricists, or at least their views can easily be adapted along these lines (Carus 2007, 1–40, 297–309; 2017; Friedman 2001, 93–103;

Uebel 2004, 2020).[3] Although these two interpretations are certainly different, since the first completely rejects the logical empiricist conception of values, while the second defends it, albeit with caveats, both interpretations have in common that the interpreters seem to think that there is a "relativist" potential in logical empiricism that needs to be corrected towards a more cognitive understanding of deliberation. To illustrate the difference between these third-generation interpretations and the views held by fourth-generation scholars today—I rely here mainly on verbal discussions—is that most scholars today seem to think that such a correction is not necessary at all.

3. *Other currents either reject noncognitivism or science or both.* Two important debates must be mentioned.

   3a. The debate with the *Frankfurt School*, which led to a wild polemics (Horkheimer 1937) as well as a more moderate account (Horkheimer and Adorno 1969). Horkheimer and Adorno fundamentally criticized the scientific worldview as a whole—science needed to be controlled from the outside by philosophers being trained in dialectics—and countered noncognitivism with cognitivism regarding values, again based on dialectics. Politics, according to the Frankfurt School, hardly needs science, but absolute values being contributed by wise philosophers. This picture remained the lore of Frankfurt School up to the positivism struggle of the 1960s (Dahms 1994, Zweiter Teil), and it was only in the 1970s that Jürgen Habermas started to develop a much more refined view, partly in cooperation with advocates of analytic philosophy and American pragmatism. Habermas was no longer opposed to science, although, lacking a clear understanding of instrumental rationality, his conception never makes it entirely clear what role science can play in politics; and he replaced the academic prophecy of Adorno and Horkheimer with a deliberativist view, according to which values are no longer to be identified by experts but in a careful process of cognitive deliberation involving all rational citizens. The result is a consensus that represents the objective values of the state (up to the next election); those who do not share the consensus must accept that they act irrational (Habermas 1992, 2022).

   3b. The second major debate in which logical empiricism has been involved since the 1930s is the debate with *American pragmatism*. Similar to the case of the Frankfurt School, promising beginnings were quickly followed by the end of the debate (Richardson 2003; Limbeck-Lilienau 2010). The pragmatists theory of value deliberation was developed, with the exception of (Dewey 1939), with little interaction with logical empiricism, though with a thoroughly critical attitude toward the latter (Lewis 1946). The American pragmatists always supported the scientific worldview in much the same

[3] My own view is close to, though not identical with, the second interpretation, and it obviously contradicts the first interpretation. A more detailed account of the relationship between the interpretation presented here and these two interpretations has to be postponed here.

way as the logical empiricists, i.e., they also supported the instrumental rationality of rules (1) through (4) as outlined in the previous section. However, they shared with the Frankfurt School a rejection of noncognitivism and thus defended a cognitive naturalism with respect to values. More recently, Cheryl Misak has argued convincingly that this view fits well with the mature views of Habermas, and that both Lewis and Habermas can be taken as good foundations for a cognitive theory of deliberation (Misak 2000).

4.6 Concluding Remark: Noncognitive Deliberation Today

Noncognitive deliberation has little to do with meta-ethics, contrary to the still dominant view that Carnap's and other logical empiricists' views on values were meta-ethical in nature. That values are noncognitive and cannot be objectively discovered by means of some Platonic heaven is a triviality to which the views of the logical empiricists can by no means be reduced. Rather, the logical empiricist view represents a theory of moral and political deliberation, which differs from the standard picture (Bächtiger et al. 2018) in two ways. First, it puts more emphasis on rationality and science as a means of deliberation. Second, it rejects the idea that deliberation must always reach a universal consensus, replacing it with the Kelsenian notion of compromise and the possibility of productive "agonism".[4]

References

Ayer, Alfred Jules. 1946. *Language, Truth and Logic*. New York: Dover Publications.

Bächtiger, André, John S. Dryzek, Jane Mansbridge, and Mark E. Warren, eds. 2018. *The Oxford Handbook of Deliberative Democracy*. Oxford: Oxford University Press.

Carnap, Rudolf. 1932. Überwindung der Metaphysik durch logische Analyse der Sprache. *Erkenntnis* 2: 219–241.

———. 1934. Theoretische Fragen und praktische Entscheidungen. *Natur und Geist* 2: 257–260.

———. 1935. *Philosophy and Logical Syntax*. London: Kegan Paul.

———. 1937. Logic. In *Factors Determining Human Behavior*, ed. Edgar Douglas et al., 107–118. Cambridge, MA: Harvard University Press.

———. 1944. Note on Values. In *Ray Lepley: Verifiability of Value*, 137–138. New York: Columbia University Press.

———. 1950. Empiricism, Semantics and Ontology. *Revue International de Philosophie* 4: 20–40.

———. 1963a. Intellectual Autobiography. In *The Philosophy of Rudolf Carnap*, ed. Paul Arthur Schilpp, 3–84. La Salle: Open Court.

———. 1963b. Replies and Systematic Expositions. In *The Philosophy of Rudolf Carnap*, ed. Paul Arthur Schilpp, 859–1013. La Salle: Open Court.

———. 1998 [1928]. *Der logische Aufbau der Welt*. 3rd ed. Hamburg: Felix Meiner Verlag.

[4] This view is developed in more detail in Christian Damböck, "Noncognitive Deliberation. The Political Legacy of Logical Empiricism", *Erkenntnis*, 2024.

———. 2017. Value Concepts (1958). *Synthese* 194: 185–194.

Carnap, Rudolf, and Willy Hochkeppel. 1967. Andere Seiten des Denkens. Willy Hochkeppel sprach mit Rudolf Carnap. *Der Monat* 19 (224): 50–56.

Carus, A.W. 2007. *Carnap and Twentieth-Century Thought. Explication as Enlightenment*. Cambridge: Cambridge University Press.

———. 2017. Carnapian Rationality. *Synthese* 194: 163–184.

———. 2021. Werte beim frühen Carnap: Von den Anfängen bis zum *Aufbau*. In *Der junge Carnap in historischem Kontext*, ed. Christian Damböck and Gereon Wolters, 1–18. Cham: Springer.

———. 2022a. Carnap's Fundamental Philosophical Commitment: From Religious Origins to Kantian Non-cognitivism, 1911–21. In *The Vienna Circle and Religion*, ed. Esther Ramharter, 3–22. Cham: Springer.

———. 2022b. Die religiösen Ursprünge des Nonkognitivismus bei Carnap. In *Logical Empiricism, Life Reform, and the German Youth Movement/Logischer Empirismus, Lebensreform und die deutsche Jugendbewegung*, ed. Christian Damböck, Günther Sandner, and Meike Werner, 143–162. Dordrecht: Springer.

Dahms, Hans-Joachim. 1994. *Positivismusstreit. Die Auseinandersetzung der Frankfurter Schule mit dem logischen Positivismus, dem amerikanischen Pragmatismus und dem kritischen Rationalismus*. Frankfurt/Main: Suhrkamp.

———. 2004. *Neue Sachlichkeit* in the Architecture and Philosophy of the 1920s. In *Carnap Brought Home. The View from Jena*, ed. Steve Awodey and Carsten Klein, 357–376. Chicago: Open Court.

———. 2021. Rudolf Carnap: Philosoph der Neuen Sachlichkeit. In *Der junge Carnap in historischem Kontext: 1918–1935*, Hrsg. Christian Damböck und Gereon Wolters, 75–106. Cham: Springer.

Damböck, Christian. 2017. *<Deutscher Empirismus>. Studien zur Philosophie im deutschsprachigen Raum 1830–1930.Veröffentlichungen des Instituts Wiener Kreis*. Dordrecht: Springer.

———. 2018. Die Entwicklung von Carnaps Antimetaphysik, vor und nach der Emigration. In *Deutschsprachige Philosophie im amerikanischen Exil 1933–1945. Historische Erfahrung und begriffliche Transformation*, Hrsg. Max Beck und Nicholas Coomann, 37–60. Vienna.

———. 2022a. Carnap's Non-cognitivism and His Views on Religion, Against the Background of the Herbartian Philosophy of his Grandfather Friedrich Wilhelm Dörpfeld. In *The Vienna Circle and Religion*, ed. Esther Ramharter, 23–40. Dordrecht: Springer.

———. 2022b. Carnap, Reichenbach, Freyer. Non-cognitivist Ethics and Politics in the Spirit of the German Youth-Movement. In *Logical Empiricism, Life Reform, and the German Youth-Movement*, ed. Christian Damböck, Günther Sandner, and Meike Werner, 163–180. Cham: Springer.

———. 2022c. Einleitung. In *Rudolf Carnap, Tagebücher 1908–1919. Herausgegeben von Christian Damböck*, 9–58. Hamburg: Meiner.

———. 2022d. Nachwort. In *Rudolf Carnap: Überwindung der Metaphysik durch logische Analyse der Sprache*, 63–100. Ditzingen: Reclam.

———. 2022e. The Politics of Carnap's Non-Cognitivism and the Scientific World-Conception of Left-Wing Logical Empiricism. *Perspectives on Science* 30 (4): 493–524. https://doi.org/10.1162/posc_a_00372.

———. 2024a. Carnap's Non-Cognitivism: Paths and Influences. In *Interpreting Carnap*, Hrsg. Alan Richardson und Adam Tamas Tuboly, 13–31. Cambridge: Cambridge University Press.

———. 2024b. Carnaps Ästhetik: *Der logische Aufbau der Welt* im Kontext des Nach-Expressionismus. In *Franz Roh 1890-1965. Eine Kollektivbiografie*, ed. Hans-Joachim Dahms. Wien: Virtual Archive of Logical Empiricism, 83-108.

———. 2024c. Entscheidungstheoretische, modallogische und metaethische Gesichtspunkte in Carnaps 'Value Concepts (1958)'. In *Revisiting the History and Philosophy of Logic and Mathematics*, ed. Elena Ficara, Julia Franke-Reddig, Anna-Sophie Heinemann, and Andrea Reichenberger. College Publications: forthcoming.

Danneberg, Lutz, Andreas Kamlah, and Lothar Schäfer, eds. 1994. *Hans Reichenbach und die Berliner Gruppe*. Edited by Lutz Danneberg and Andreas Kamlah. Braunschweig: Vieweg.
Daston, Lorraine. 2001. *Wunder, Beweise und Tatsachen. Zur Geschichte der Rationalität*. Frankfurt/Main: Fischer.
Dewey, John. 1939. *Theory of Valuation*. Chicago: Chicago University Press.
Dewulf, Fons. 2022. Leo Apostel and Rudolf Carnap: The Development of Logical Empiricist Ethics in Post-War Europe. In *The Socio-Ethical Dimension of Knowledge*, ed. Christian Damböck and Adam Tuboly, 127–147. Cham: Springer.
Friedman, Michael. 2001. *Dynamics of Reason. The 1999 Kant Lectures at Stanford University*. Stanford: Center for the Study of Language and Information.
Habermas, Jürgen. 1992. *Faktizität und Geltung. Beiträge zur Diskurstheorie des Rechts und des demokratischen Rechtsstaats*. Frankfurt/Main: Suhrkamp.
———. 2022. *Ein neuer Strukturwandel der Öffentlichkeit und die deliberative Politik*. Berlin: Suhrkamp Verlag.
Hegselmann, Rainer. 1979. *Normativität und Rationalität. Zum Problem praktischer Vernunft in der analytischen Philosophie*. Frankfurt/Main: Campus Verlag.
Heidelberger, Michael. 2024. Between Pietism and Herbartianism: Archaeological Vestiges in Carnap's Thought. In *Ways of the Scientific World Conception. Rudolf Carnap and Otto Neurath*, ed. Christian Damböck, Johannes Friedl, and Ulf Höfer. Leiden: Brill: forthcoming.
Horkheimer, Max. 1937. Der neueste Angriff auf die Metaphysik. *Zeitschrift für Sozialforschung* 6 (1): 4–51.
Horkheimer, Max, and Theodor W. Adorno. 1969. *Dialektik der Aufklärung. Philosophische Fragmente*. Frankfurt/Main: Fischer.
Kamlah, Andreas. 1977. Erläuterungen zu Kapitel 17: Der Ursprung von Reichenbachs Nonkognitivismus in der Jugendbewegung. In *Hans Reichenbach: Der Aufstieg der wissenschaftlichen Philosophie. Mit einer Einleitung zur Gesamgausgabe von Wesley C. Salmon und mit Erläuterungen von Andreas Kamlah*, 480–483. Wiesbaden: Springer Fachmedien.
Kelsen, Hans. 1929. *Vom Wesen und Wert der Demokratie. Zweite, umgearbeitete Auflage*. Tübingen: J.C.B. Mohr (Paul Siebeck).
———. 1932. Verteidigung der Demokratie. *Blätter der Staatspartei* 2: 90–98.
———. 1937. Wissenschaft und Demokratie. *Neue Zürcher Zeitung* 321 (February 23): 1–2.
———. 1955. Foundations of Democracy. *Ethics* 66: 1–101.
Kraus, Oskar. 1937. *Die Werttheorien. Geschichte und Kritik*. Brünn: Verlag Rudolf M. Rohrer.
Kuhn, Thomas S. 1996. *The Structure of Scientific Revolutions*. 3rd ed. Chicago: The University of Chicago Press. 1962.
Lewis, Clarence Irving. 1946. *An Analysis of Knowledge and Valuation*. La Salle IL: Open Court.
Limbeck-Lilienau, Christoph. 2010. Rudolf Carnap und die Philosophie in Amerika. Logischer Empirismus, Pragmatismus, Realismus. In *Vertreibung, Transformation und Rückkehr der Wissenschaftstheorie. Am Beispiel von Rudolf Carnap und Wolfgang Stegmüller*, ed. Friedrich Stadler, 85–164. Berlin: Lit Verlag.
Linsbichler, Alexander. in preparation. The Fact/Value Dichotomy as an Alleged Thought Stopper. In *Carnap Handbuch*, ed. Christian Damböck and Georg Schiemer. Stuttgart: Metzler Verlag.
Linsbichler, Alexander, and Ivan Ferreira da Cunha. 2023. Otto Neurath's Scientific Utopianism Revisited – A Refined Model for Utopias in Thought Experiments. *Journal for General Philosophy of Science*. https://doi.org/10.1007/s10838-022-09630-5.
Mach, Verein Ernst, ed. 2012 [1929]. *Wissenschaftliche Weltauffassung. Der Wiener Kreis. Herausgegeben von Friedrich Stadler und Thomas Uebel*. Vienna: Springer-Verlag.
Misak, Cheryl. 2000. *Truth, Politics, Morality. Pragmatism and Deliberation*. London: Routledge.
Mormann, Thomas. 2006. Werte bei Carnap. *Zeitschrift für philosophische Forschung* 60: 169–189.
———. 2007. Carnap's Logical Empiricism, Values, and American Pragmatism. *Journal for General Philosophy of Science* 38: 127–146.
———. 2010. Wertphilosophische Abschweifungen eines Logischen Empiristen: Der Fall Carnap. In *Logischer Empirismus, Werte und Moral: Eine Neubewertung*, Hrsg. Anne Siegetsleitner, 81–102. Wien: Springer.

———. 2016. Carnap's *Aufbau* in the Weimar Context. In *Influences on the Aufbau. Vienna Circle Institute Yearbook*, ed. Christian Damböck, 115–136. Dordrecht: Springer.
Mouffe, Chantal. 1993. *The Return of the Political*. London: Verso.
Neurath, Otto. 1981. *Gesammelte philosophische und methodologische Schriften*. Wien: Hölder-Pichler-Tempsky.
———. 1996. Visual Education. Humanisation versus Popularisation. In *Encyclopedia and Utopia. The Life and Work of Otto Neurath (1882–1945)*, ed. Elisabeth Nemeth and Friedrich Stadler, 245–336. Dordrecht: Kluwer Academic Publishers.
Putnam, Hilary. 1983. *Realism and Reason. Philosophical Papers, Volume 3*. Cambridge: Cambridge University Press.
———. 2002. *The Collapse of the Fact/Value Dichotomy and Other Essays*. Cambridge, MA: Harvard University Press.
Reichenbach, Hans. 1951. *The Rise of Scientific Philosophy*. Berkeley: University of California Press.
———. 1978 [1913]. The Free Student Idea: Its Unified Content. In *Hans Reichenbach: Selected Writings 1909–1953. Volume One*, 108–123. Dordrecht: D. Reidel Publishing Company.
Reisch, George A. 2005. *How the Cold War Transformed Philosophy of Science. To the Icy Slopes of Logic*. Cambridge: Cambridge University Press.
Richardson, Alan. 1998. *Carnap's Construction of the World. The Aufbau and the Emergence of Logical Empiricism*. Cambridge: Cambridge University Press.
———. 2003. Logical Empiricism, American Pragmatism, and the Fate of Scientific Philosophy in North America. In *Logical Empiricism in North America*, ed. Gary Hardcastle and Alan Richardson, 1–24. Minneapolis: University of Minnesota Press.
———. 2007. Carnapian Pragmatism. In *The Cambridge Companion to Carnap*, ed. Michael Friedman and Richard Creath, 295–315. Cambridge: Cambridge University Press.
Romizi, Donata. 2009. The Vienna Circle's 'Scientific World Conception': Philosophy of Science in a Political Arena. *HOPOS* 2: 205–242.
Schlick, Moritz. 2006. *Lebensweisheit. Versuch einer Glückseligkeitslehre. Fragen der Ethik. Herausgegeben und eingeleitet von Mathias Iven. (Moritz Schlick Gesamtausgabe. Abteilung I, Band 3)*. Wien: Springer-Verlag.
Siegetsleitner, Anne, ed. 2010. *Logischer Empirismus, Werte und Moral. Eine Neubewertung*. Wien: Springer.
———. 2014. *Ethik und Moral im Wiener Kreis. Zur Geschichte eines engagierten Humanismus*. Wien: Böhlau.
Simmel, Georg. 1913. Das individuelle Gesetz: ein Versuch über das Prinzip der Ethik. *Logos* 2: 117–160.
Stadler, Friedrich. 2015. *The Vienna Circle: Studies in the Origins, Development, and Influence of Logical Empiricism*. Dordrecht: Springer.
Stuchlik, Joshua. 2011. Neurath's Naturalization of Ethics. *HOPOS: The Journal of the International Society for the History of Philosophy of Science* 1 (2): 183–208.
Uebel, Thomas. 2004. Carnap, the Left Vienna Circle, and Neopositivist Antimetaphysics. In *Carnap Brought Home. The View from Jena*, ed. Steve Awodey and Carsten Klein, 247–278. Chicago: Open Court.
———. 2020. Intersubjective Accountability: Politics and Philosophy in the Left Vienna Circle. *Perspectives on Science* 28 (1): 35–62.
van Fraassen, Bas C. 1980. *The Scientific Image*. New York: Oxford University Press.
Werner, Meike G. 2022. Youth and Politics at the End of the Great War: Rudolf Carnap's *Politische Rundbriefe* of 1918. In *Logical Empiricism, Life Reform, and the German Youth Movement/ Logischer Empirismus, Lebensreform und die deutsche Jugendbewegung*, ed. Christian Damböck, Günther Sandner, and Meike Werner, 105–126. Dordrecht: Springer.
Windelband, Wilhelm. 1921. *Präludien. Aufsätze und Reden zur Philosophie und ihrer Geschichte (Zwei Bände in einem)*. 8th ed. Tübingen: J. C. B. Mohr (Paul Siebeck). 1884.
Yap, Audrey. 2024. Philosophy in a Collective Spirit as "Politics in Its Broadest Sense". In *Interpreting Carnap*, Hrsg. Alan Richardson und Adam Tamas Tuboly, 70–85. Cambridge: Cambridge University Press.

Chapter 5
Scientific Humanism in Practice—L. Susan Stebbing and Otto Neurath

Silke Körber

Abstract At a time of crisis and war in Europe, the Vienna Circle member Otto Neurath and the British philosopher L. Susan Stebbing corresponded and co-operated for over a decade from the 1930s onwards. Despite programmatic differences, they agreed that the role of philosophy, science and scientific expertise needed to be reconsidered and that the formation of modern democracies and an egalitarian global society had to be actively promoted that way. The aim was to improve the lives of as many people as possible and contribute to their happiness. For this purpose, it was necessary to find new techniques of knowledge transfer for the general public, but also to train people in critical thinking and in the judgement of facts and data—as an independent, questioning attitude dealing with institutions, politics, or the media. The requirements for transfer on the knowledge production side had to be reflected accordingly. The modernity of Neurath and Stebbing's respective thoughts lies in their focus on the ordinary use and transfer of knowledge, in questioning the unlimited authority of scientific experts as well as the ideal of science at the time—in favour of a more pluralistic approach to fight propaganda and disinformation and support social change.

Keywords L. Susan Stebbing · Otto Neurath · Scientific Humanism · Unity of Science · Popular Education and Knowledge Transfer · Critical Thinking · Common Sense · Scientific Experts · Vienna Circle · Cambridge School of Analytic Philosophy

---

This article is in part based on an analysis of Stebbing and Neurath's correspondence and cooperation "Thinking Clearly for a Modern Democracy Susan Stebbing and Otto Neurath" that I co-authored with Adam Tuboly in a forthcoming volume on Susan Stebbing, edited by Siobhan Chapman and Brian Garvey (Palgrave Macmillan, 2024/5).

S. Körber (✉)
Humboldt University, Berlin, Germany
e-mail: silke.koerber@alumni.hu-berlin.de

G. Schiemer (ed.), *The Legacy of the Vienna Circle*, Vienna Circle Institute Yearbook 30, https://doi.org/10.1007/978-3-031-80568-4_5

5.1 Science for a Better Life

In the first half of the twentieth century, various actors and groups in society in general as well as in academia discussed the relation between scientific and general knowledge as well as the prerequisites and appropriate forms of communication. Central themes of modernity emerged in these discourses, were discussed and led to different programmatic consequences: from influences of the new (also visual) mass media, the triumph and crisis of individualism, the longing for an old order and hopes or doubts about the latest achievements of science and technology. The latter had permanently changed Europe in the nineteenth century, the "scientific age", as Werner von Siemens had called it. In this context, the ongoing process of professionalisation, standardisation and specialisation in the sciences was significant—and, at the time, promoted both the crisis and reorganisation of the academic system itself. The enormous technical progress and gain in knowledge in the "positivist" sciences, from the natural to the technical sciences, especially physics with quantum mechanics and Einstein's theory of relativity, and applied mathematics, was not matched by comparable achievements in the humanities. This applied in particular to philosophy, understood by Immanuel Kant and Wilhelm von Humboldt as the leading discipline or foundation of the sciences, and the associated neo-humanist ideal of education. But the accessibility and usability of scientific findings in the technical and natural sciences was (like in the humanities) still limited for certain social groups against the background of very different levels of education, availability of information and the lack of transparency of the latter.

At the same time, access to increasingly specialised knowledge in European cities from London to Vienna became a prerequisite for participation and both social and economic advancement. It constituted the broader context for the positioning of the Vienna Circle and of a "scientific worldview" as well as for the further development of analytical philosophy, before and after the Second World War. In this context, the collaboration between the British philosopher L. Susan Stebbing (1885–1943) and the Vienna Circle member Otto Neurath (1882–1945) could be seen as an alliance that approached central problems from different but compatible angles. Both addressed developments of modern societies in their theoretical and scientific work as well as their public activities and in more popular publications.[1] Even though their correspondence and their decade long collaboration has received little attention so far, it allows to further contextualise some of the debates at the time. Furthermore, it is relevant for current discourse on the role of science and experts as advisors in times of crises and on the preconditions and practise of knowledge transfer to the general public. Both, Stebbing and Neurath, positioned themselves against a one-sided, traditional understanding of scientific knowledge and education marked by social, cultural or religious boundaries. Both called for a respectful, serious approach to the transmission of information in the steps of

[1] See for Neurath e.g. Cat and Tuboly (2019), for Stebbing Chapman (2013), Körber (2019, 2023), and Janssen-Lauret (2022).

selection, analysis, and a constant revision of information to strengthen the general ability of reasoning and judgement. Thus, the knowledge to be conveyed—given the processual, non-final character of science—was to be understood less as a secure, validated corpus that must be communicated to an "ignorant layperson". Rather, it was about the assessments of a topic or subject area by a group of experts in a specific epistemic environment for other scientists and society in general. In exile, Neurath seems to have broadened the scope of his educational work again to foster a general improvement of life for as many people as possible and an international community—in terms of promoting the ability to analyse and judge situations and things in the world, which is part of a comprehensive upbringing in democratic societies. This, like Stebbing, seems to him to be a prerequisite for a fulfilled, happy life of the average citizen in the community and for ensuring equality and participation—and eventually peace. For the socialist Neurath, as well as for the British intellectual and socially committed philosopher Stebbing, scientists (and philosophers) should support the collective: not by making decisions, either in politics and other institutions, but by helping to create and promote a basis for decision-making, through the transfer of information, the training and development of the citizens' ability to judge, and by the promotion of scientific progress as a whole. However, this should not be misunderstood as an uncritical, unreflective adoption of scientific research or the statements of scientific experts in general. Like Neurath (e.g. in "Visual Education: Humanisation versus Popularisation", 1945), Stebbing in particular proved to be a harsh critic of supposedly popular presentations of current scientific findings (e.g. in *Philosophy and the Physicists*, 1937) or political, economic and social issues for a broad audience (*Thinking to Some Purpose*, 1939b).

The focus of this text is therefore on the one hand on the presentation of Neurath's and Stebbing's central positions with regard to the goals of humanistic and enlightened thinking and the role of scientific knowledge and experts in democratic societies. On the other hand, some of the main differences between the British philosopher and the Vienna Circle member concerning means and methodology of knowledge dissemination will be addressed. In order to make this comprehensible, the most important information about Susan Stebbing, Otto Neurath's activities in exile and the framework and character of their co-operation and communication will be briefly explained. What follows is a description of her central reflections on the task of science and the scientist. Finally, the differences are discussed, which did not lead to a divide, but to complementing positions—even if the early death of both, a lack of students and official schools, certainly prevented an expansion or, on Neurath's part, the continuation of the work they had begun together.

5.2 Neurath and Stebbing—Towards Scientific Humanism

The correspondence between Otto Neurath and L. Susan Stebbing is part of the Vienna Circle collection and the Otto Neurath holdings contained therein, which are located in the Noord Hollands Archief in Haarlem. Their exchange of letters began during his exile in The Hague and stopped with his flight to England and then finally with the end of his internment on the Isle of Man in early 1941, with Stebbing supporting him and his later wife Marie Reidemeister in the process. Neurath and Reidemeister settled with the ISOTYPE Institute (International System of Typographic Picture Education) in Oxford and direct personal contact now replaced the correspondence with Stebbing. Their growing friendship and collaboration lasted until Stebbing's early death in 1943 from cancer.

On 30 July 1934 Neurath wrote a letter to Stebbing in London to invite her to the First International Congress for the Unity of Science in Paris in 1935 and to a preparatory meeting in Marienbad in August 1934, which was to take place shortly before the International Congress of Philosophy in Prague. This formal invitation marked the beginning of a regular exchange of letters. Stebbing became influential for Neurath in many ways, first as a member of the Unity of science movement and organizer of their conference in the UK (Cambridge 1938). She became also a member of the Board of Directors of the International Foundation for Visual Education, that Otto and his team founded in The Hague—as well as the first chairperson of the ISOTYPE Institute in Oxford. During Neurath's exile they did not only communicate and support each other's work. They read each other's publications, as is evident from their correspondence, references in texts[2] and also from Otto Neurath's exile library. He kept copys of her publications, made annotations and highlighted passages he read, e.g. in her last publication *Men and Moral Principles* (1944) (see Sect. 11.3 in this volume and in the manuscript "The Inadequacies of the Scientific Attitude", see Sect. 11.2). In 1938 Neurath wrote in a letter to Stebbing: "[...] you are flesh of the flesh of Logical Empiricism."[3] And in fact, she was already closely connected to the "scientific worldview" and the Vienna Circle even before they established contact: she knew and corresponded with the physicist and philosopher Moritz Schlick since 1930. She invited Rudolf Carnap to give a lecture in London in 1934, then also others, and became well acquainted with the various positions of the group and the analytic method, even before Alfred Ayer helped to introduce the movement to the English-speaking world. However, this did not mean that Stebbing simply joined Neurath's activities in line with the Vienna Circle's programme. When he established contact with her, she was already a renowned professor in London with an international reputation and a public intellectual in the UK. Against

[2] In *Modern Man in the Making*, Otto Neurath had already referred to Stebbing's Thinking to Some Purpose in the annotations (Neurath 1939, 136). And as the correspondence shows, he had been reading her book with great interest: I read your blue booklet [Thinking to Some Purpose] through and was very pleased to find the remarks about Visual Miseducation. Excuse me, please, that I beg you to write a review on my booklet[s] International Picture Language. The first rules of ISOTYPE and Basic by ISOTYPE. I seek always a reviewer for these books, but nobody of our people who are interested in Logic are interested in Visual Education and ISOTYPE too. You are the first; I am very glad that you are full of educational ideas. It is nice that you are a member of the board of directors of our International Foundation of Visual Education. See Neurath to Stebbing, April 8, 1939, Otto Neurath's Nachlass (ONN 400/R.76) at Wiener Kreis Archief, Noord-Hollands Archief, Haarlem).

[3] Otto Neurath in a letter to L. Susan Stebbing of 12 February 1938 (ONN).

the background of political crisis and war she became increasingly concerned with communicating knowledge to a general public in writings and lectures. In order to fully understand and analyse these contexts, it is necessary to provide a brief overview of main biographical stages in Stebbing's life, which are of particular significance.[4]

Born in 1885, Stebbing was able to study History at Girton College in Cambridge from 1904 to 1908—which was an exception at that time for women. As Janssen-Lauret phrased it: to even attend a woman's college was an act of emancipation to "rise through the ranks as an academic even more so" (Janssen-Lauret 2022, 1). Following her interest, Stebbing went on to study Philosophy, graduating with a 'Master of Moral Science' from Kings College in 1912. In 1933, she became the first woman in the UK to be appointed Professor of Philosophy at Bedford College in London. Stebbing was co-founder of the journal *Analysis*, chairwoman of the *Aristotelian Society* (1933/34) and president of the *Mind Association* (1931/32), a honorary associate of the Rationalist Press Association, and a supporter of the League of Nations Union. During her studies, she was, among others, influenced by G. E. Moore's analytical and common-sense philosophy. She was particularly interested in modern logic, philosophy of science and philosophy of language as also represented by Bertrand Russell and Alfred N. Whitehead. In summary, Stebbing can be regarded as a proponent of what later became known as the Cambridge School within analytic philosophy, which was committed to logical analysis and also to the interpretation of the "common language" for solving scientific problems. In her standard work *A Modern Introduction to Logic* (1930), modern mathematical logic was treated on an equal footing with Aristotelian logic. According to more recent research, Stebbing was the only woman among the ten the most frequently cited authors in four British philosophy journals (e.g. *Philosophy*, Analysis, *Mind*) in from 1931–1940[5] and influenced students like Max Black (1909–1988) or Margaret MacDonald[6] (1907–1956). According to the Polish logician Maria Ossowska, who lived in London for 3 years during the 1930s, Stebbing assembled an illustrious group of philosophers for regular meetings: "I attend private meetings of Prof. L.S. Stebbing, around whom the most interesting and most reasonable people gather. This group publishes the journal *Analysis* …".[7] Stebbing endeavoured to put her philosophical work and humanist principles not only into theory, but also into practice, supporting refugees at her own expense and helping other scientists who fled to England during the war to gain a foothold there, taking part in anti-war activism, e.g. on behalf of the League of Nations Union. With her friends and sister,

[4]The biographical information on Stebbing is essentially based on Chapman (2013), Beaney (2016), and Janssen-Lauret (2022).

[5]This included philosophers only active in the twentieth century like Bertrand Russell and Rudolf Carnap, see Verhaegh, Petrovich, Bös (2024). The American Reception of Logical Empiricism: A Mention-Based Bibliometric Analysis (in this volume, Chap. 7).

[6]See esp. Vlasits (2022).

[7]Archiwum Kazimierza Twardowskiego, AKT K-02-1-27, Cards 42–43 (25.11.1934), Translation by Anna Brożek, Warsaw 2024.

she took in Jewish refugee children at the girls' school, Kingsley Lodge, that they had set up together.[8] And in 1941 Stebbing became the first female president of the Union of Ethical Societies, the forerunner of Humanists UK, the British association for the promotion of humanism, which campaigns until this day for secularisation, human rights, democracy, and equality. In her opinion, not religion, but philosophy and the sciences should provide orientation and help to answer the question of a "worthwhile life", as Stebbing expressed it 1939 in "Ethics and Materialism":

> Some of the unreligious, for whom materialism is assumed to be enough, are not entirely without hope of a better world. But their hopes are concerned with this world, which can be made better worth living in by the efforts of human beings. Their new Jerusalem is to be attained, if at all, in 'England's green and pleasant land' (or in Germany, etc.); it is not to be postponed to a superhistorical goal [...] and offered to those who cannot read the symbolism of mathematics. (Stebbing 1939a, 44).

This strong and comprehensive humanist approach, which Stebbing and Neurath shared, was in general rooted in open-minded, humane thinking that could be traced back to the ancient world—from China and India to Athens or Rome, the Italian Renaissance, or the rise of reason and scientific thinking during the Enlightenment. Nevertheless, the humanist thinking of the twentieth century was especially connected to the foundation of international co-operative institutions like the UNESCO and to the enormous advancements of contemporary science as an instrument as well as a guide for modern life. Alongside the Vienna Circle, Otto Neurath in particular was involved early on with humanist and internationalist projects and approaches. This context is already well known from research on Otto Neurath's educational work transforming facts and data from Economy and Sociology, Medicine or Biology into pictorial statistics at the Gesellschafts- und Wirtschaftsmuseum (Museum for Social and Economic Affairs) in Vienna between 1924/5 and 1934 and later with the ISOTYPE Institute in the Netherlands (1934–1940) and in the UK from 1941 until his early death 1945 in Oxford. He co-operated with the Belgian pioneer of information science, Paul Otlet, in the service of cultural internationalism as part of the "Mundaneum Project", an archive and "encyclopaedic" museum of world knowledge. They first met in 1929 and shared the view that a globally organised, "enlightened" knowledge community could help to secure world peace—in view of the massive social upheaval and destruction caused by the First World War. As Nader Vossoughian and Wouter Van Acker have pointed out, Otto Neurath's work took on a new, more global communication-oriented dimension through his cooperation with Paul Otlet and other representatives of cultural internationalism and encyclopaedism, moving away from the original focus on workers' and popular education in "Red Vienna", which was further encouraged by the massive political upheavals of the time, war, flight and emigration and the resulting personal situation. Neurath was motivated "[...] to engage problems of communication posed by international bodies and institutions" (Vossoughian 2003, 93).

---

[8] See Chapman (2013) and Janssen-Lauret (2022, 11).

Stebbing agreed with Neurath that changes in the scientific and philosophical worldview should have direct effects on our everyday lives, as a starting point for a reform of society and politics. As Anne Siegetsleitner described it, this kind of "scientific humanism" was already a common frame of reference within the Vienna Circle, that was at the same time open to specific interpretations (Siegetsleitner 2014, 2015). In order to understand this claim and contextualise Neurath and Stebbing's position, it is helpful to remember the main criteria of Rudolf Carnap's description of "scientific humanism" from his Autobiography of 1963: 1. "[...] man has no supernatural protectors or enemies [...] therefore whatever can be done to improve life is the task of man himself; 2. [...] mankind is able to change the conditions of life in such a way that many of the sufferings of today may be avoided and that the external and internal situation of life for the individual, the community, and finally for humanity will be essentially improved; 3. [...] all deliberate action presupposes knowledge of the world, that the scientific method is the best method of acquiring knowledge and that therefore science must be regarded as one of the most valuable instruments for the improvement of life." (Carnap 1963, 83).

As is well known, the task of philosophy was limited by the Vienna Circle to a method and a theory of science, and rejected as a discipline above and guiding the sciences. The basis for knowledge was to be modern logic on the one hand, and empirical inquiries on the other. Apart from metaethics there was no room for rationality in moral matters, meaning that there are no factual moral statements. The general programmatic orientation as set out in the manifesto *Wissenschaftliche Weltauffassung* (1929) however, took different shapes and forms: a stronger socialist embedding and reformist functions represented by the so called left wing, especially Neurath and Rudolf Carnap (Stadler & Uebel 2012). And that of the "right wing" with Moritz Schlick and Friedrich Waismann, who wanted to limit themselves primarily to a reform of the sciences and a revised, but not completely obsolete role for philosophy. Since the basic problems were identified as arising from inadequate linguistic formulation and an insufficient separation between the syntax and logic of scientific factual statements, the solution seemed to lie in the development of a suitable analytical set of rules and a more precise linguistic expression. According to Otto Neurath, this "scientification" of the form of scientific sentences should also correspond to a precise, clear and factual form of everyday information and communication, especially on social topics, but also on natural sciences and technology. Neurath's concepts of a" visual auxiliary language" ISOTYPE, and a" picture-text style" derived from it, can only be understood in the context of his preoccupation with modern symbolic languages and the utopia of a unified communication that transcends social and geographical boundaries for the purpose of disseminating modern scientific knowledge. It was the impact of enlightened science on society, mediated through various institutions and practices, as set out in the Vienna Circle's manifesto of 1929, that became a central component of the movement's programme. Neurath and Stebbing's cooperation may be seen from a very similar point of view. Here it was especially the conception of the socio-political role of the scientist and the relevance of critical analytical thinking that united them. However, in both cases this was linked to a clear division of responsibilities between citizens and political institutions on the one hand and scientific experts and the role of science in general on the other, which will be discussed in more detail below. And despite programmatic differences—which had also been present within the Vienna Circle—the means and methods that Stebbing and Neurath proposed for a better transfer of knowledge were rather stimulatingly complementary.

5.3 The "Scientific Attitude" and the Scientific Expert

Both, Neurath and Stebbing, were in fact not just concerned with the adequate transfer and formalisation of scientific knowledge, but felt the need to introduce the same standards to ordinary language and make scientific language accessible to the common people. But it seems that this has only recently been addressed or properly acknowledged in Stebbing's case (see e.g. West 2022). Louis Doulas points out that Stebbing did not only predate the late Wittgenstein and J.L. Austin in her emphasis on the analysis of "language of common sense i.e. the ordinary language" but that the context in which she uses it is different from both of them:

> Stebbing pushes common sense and the ordinary language methodology into new territory, into the context of modern science, appealing to ordinary language not only as a way to dissolve philosophical muddles but also purported scientific muddles as well. (Doulas 2022, 20).

Stebbing focused on popular books on modern science, especially *Philosophy and the Physicists*, and strongly criticized authors such as Arthur Eddington and James Jeans, who often used emotive, 'imprecise' terminology to describe phenomena such as 'distance' as 'meaningless' or 'meaningful', or characterised the size of a planet ('the smallness of the earth') and the 'loneliness' of a star, which conveyed no new knowledge to the reader (Stebbing 1937, Chap. 1). They use the language of common sense[9] to describe specialised, highly technical scientific contexts to the common reader, but this kind of "popularization" is "a grave danger to thinking clearly" and results in "muddle" rather than understanding of scientific facts (Stebbing 1937, Chap. 1). It is an inappropriate simplification that the common reader cannot resolve because she is not familiar with the scope and context in which the term is used by experts, especially in times where "spectacular" advances in the sciences alter the current worldview. Here, according to Stebbing, knowledge is needed for orientation, to adapt and deal with the environment people are living in, not as a substitute for belief systems and religion of any kind:

> Science is not a goddess or a woman. We cannot ask science, but only scientists. Moreover, we must ask our questions of the scientist at a moment when he is in a scientific temper, capable of giving us 'the ascertained facts and provisional hypotheses' without any

[9] Common sense is to be understood as a "fragmented" and "untidy" sense of the world contrasted with the exact and orderly world of science, see Stebbing (1930), cited from Janssen-Lauret (2022, 34).

> admixture of the emotional significance which he reads into these facts in his scientific moods. (Stebbing 1937, 9)

It becomes clear that for Stebbing science as a collective endeavour is not to be mistaken with the individual scientist, who is not free of interests, values and independent of the circumstances of empirical experience—this point is being made even more explicitly in her unpublished text on the "The Inadequacies of the Scientific Attitude" (1941/2), published in this volume as well (see Sect. 11.2). In the aforementioned text "Ethics and Materialism" (1939) she criticised that so-called experts do not separate scientific statements from implicit or explicit normative ethical claims. They offer only unclear or dubious explanations, false advice and hope for a better life in the future as an excuse not change the situation for those living under difficult conditions (Stebbing 1939a, 44). In her review of the book *The Scientific Attitude* by C.H. Waddington (1941) published by Penguin in the series "Pelican Books"[10] (see Sect. 11.2 of this volume) she refers to a negative example of a "scientific attitude" which—in contrast to Neurath's and her concepts—is misunderstood as a fundamentalist approach, in which science is the only system (according to Waddington, the others are: socialism, capitalism, religion and communism) that can still be trusted to improve people's lives and benefit society:

> [...] scientists are not merely scientists; they are also human beings. Their specialist-studies provide them with special information highly relevant to the determination of what, in a given situation, we ought to do provided that we desire such & such a state of affairs. That is what scientists are often able to give us help in knowing how to achieve our ends; but the scientist – in his capacity of a scientist – does not determine those ends.[11] (Stebbing 1941/2, 9)

Stebbing—in contrast to Waddington—distinguishes between "ethical neutrality" of science and that of the "scientist". In her opinion, scientists play an immensely important role within society as providers of information on the basis of which decisions can be made for society and in the political sphere, but as individuals they are bound by socio-political, cultural or ethical contexts and convictions (Stebbing 1941/2, 9, 13). In this sense, the individual scientist must not be confused with science itself, a mistake that, according to Stebbing, is made precisely with regard to the discussion of epistemic and non-epistemic values: Waddington assumes "[...] that 'science itsef implies an ethical standard'", that science is "not neutral" (1941/2, 9), an assessment she does not share. Although the pursuit of scientific goals is in itself something good and worthy of appreciation, the values associated with it are those of individual actors, not those of science itself (1941/2, 10).

[10] The "Penguin Books", which were particularly popular during the war and in the post-war period, were described by the Guardian as a kind of "informal university" because of their compact format and generally easily accessible content (Bowler 2009, 114ff).

[11] This and following quotes are based on the handwritten undated manuscript of L.S. Stebbing, "The Inadequacies of the Scientific Attitude" from Otto Neurath's Nachlass (ONN 400/R.76), around 1941/2. For more the printed and edited version of Stebbing's text see Sect. 11.2 of this volume.

To understand Stebbing's concept for a more productive and meaningful transfer of scientific knowledge, her own bestselling book, published in 1939 as a "Pelican Book", can be consulted. *Thinking to Some Purpose* is a philosophical text directed at the general public, in which she aims to educate people about how language analysis and critical thinking can benefit society while also training its readers in how to practise these techniques. It is in fact a manual to philosophical thinking as it is described on the cover: "A manual of first-aid to clear thinking, showing how to detect illogicalities in other people's mental processes and how to avoid them in our own." (Stebbing 1939b/1959) In Stebbing's view, the principles of logical analysis are connected in the best sense with thinking "towards a purpose", as "directed" thinking, aimed at answering a question or solving a problem. The text was based on a proposal for some talks that Stebbing intended to give on the BBC radio program, but could not be carried out for some reasons. Instead she turned the manuscript into a book. Here she explains:" To think logically is to think relevantly to the purpose that initiated the thinking; all effective thinking is directed to an end. "(Stebbing 1939b/1959, 19) As in her earlier publication she distanced herself from a priori judgements, which "[...] confuses logical thinking with attempting to derive knowledge about what happens in the world by purely a priori speculations. Such an attempt is, however, thoroughly illogical; it is anti-scientific" (Stebbing 1937, 9). She illustrates her arguments with a series of analyses of contemporary texts, taken from sources such as newspapers, political speeches and advertising. Stebbing pays close attention to how language is ordinarily used in this kind of interaction. And she calls for a conscious use of language on the side of experts, whether in politics, science or the media. That means, a non-manipulative, non-emotional and non-simplifying transmission of knowledge to enable logical thinking and the formation of one's own judgement for practical democratic action. She was

> convinced of the urgent need for a democratic people to think clearly without the distortions due to unconscious bias and unrecognized ignorance. Our failures in thinking are in part due to faults which we could to some extent overcome were we to see clearly how these faults arise. (Stebbing 1939b/1959, 9)

If people had a clear and coherent view of the world with and through science, in the long term, this would lead to empowerment of citizens and more participation: For Stebbing, philosophy was not simply about solving theoretical problems. It was a way of life—and, as such, not something that required a university education or a post at a prestigious academic institution—a view she shared with Otto Neurath. Both believed that the ability to reflect on one's own motives, actions and attitudes could be taught, regardless of age, status, nationality or gender. Stebbing was less concerned with a lack of cognitive skills or of relevant information. Rather, she focused on the avoidance of errors in thinking and hasty conclusions that result, for example, from misleadingly expressed expert opinions or from the unreflected acceptance of positions by a lay audience. Thus, even in a parliamentary democracy like the British, the free formation of opinion was bound to and dependent on the availability of valid information and its critical examination. The difficult process of analyzing and evaluating complex, sometimes divergent opinions and contexts was a duty that no citizen could avoid to preserve the freedom of decision that Stebbing identified as a core element of democratic decision-making. As Frederique Janssen-Lauret has pointed out: “Political action, according to Stebbing, was not a wholly separate category from reasoning but continuous with it and dependent to a large extent on good habits of thought” (Janssen-Lauret 2022, 55).

For Stebbing, there is an obvious connection between sound, clear and logical thinking and effective action—but also a strong negative cause-and-effect relationship between unclear, “muddled” thinking and action. As Eric Schliesser pointed out recently, this correlation is anything but obvious: since biases, some form of ignorance, tradition, habits, prejudice, and many other unwanted forms of thinking and arguing still led to democratic and moral acts within society. And in our non-ideal world they play a substantial part (Schliesser 2023). Also, this rejection of “muddled thinking” was to a certain extent at odds with Neurath’s pluralistic and processual concept of science and a “universal slang” always in motion and revision, momentarily transformed into general knowledge. As explained above, for Neurath, science is a collective endeavour that must be constantly developed and revised, as is the understanding of a “scientific worldview” and the unity of science. In this sense, all sciences are connected to events and objects in the world and the individual experts are not undisputed, absolute authorities, but ultimately contribute to the development of models of interpretation and options for action to serve the welfare of humanity. On the basis of empirically validated statements in common language, the critical examination and successive acceptance of experts or scholars, their positions and ideas by the scientific community was essential. Moreover, for Neurath, an artificial language in its “pure form”, as a “metaphysics-free” and physicalist language, was ultimately only realizable in sub-areas and individual scientific disciplines. One could only successively develop a “universal slang” on the grounds of a basic vocabulary of ordinary, common language and additional scientific terms. This mixed form would have to be constantly expanded or corrected in order to keep pace with the current state of research and to enable the cooperation of scientists on an international basis, a process that could never really be completed, hence his famous ship simile. Looking at Stebbing’s criticism of the role of the scientific expert in the context of Otto Neurath’s ideas, it becomes clear that both are very close to contemporary notions of knowledge transfer: the idea of a one-sided, hierarchical understanding of the transfer of knowledge from experts to laypeople is rejected as highly simplistic in the history of ideas and science today, as the process of communication and the interplay between different actors is not sufficiently taken into account. This applies in particular to the media and education systems, which have become increasingly differentiated: traditions of knowledge accumulation and systematisation are linked to social structures, practices and techniques as well as specific interests. Even experts in certain fields can make mistakes, commit errors of reasoning, contradict each other and compete with each other, which influences their judgements and the generating of data. As Neurath claimed in his text “Visual Education: Humanisation versus Popularisation”, 1945, the common man should “have enough information to hold the power of the expert in check” (Neurath 1996, 251).

As far as Otto Neurath is concerned, there are two levels on which his activity unfolds from 1934 onwards: the scientific level, which primarily comprises the work within the framework of the Unity of Science movement, and on the other hand the broader educational activities for a general audience first in the Netherlands, then in Oxford at the ISOTYPE Institute, founded with this wife Marie after fleeing from the Nazis to the UK. The development of ISOTYPE can be seen in analogy to the development of a scientific language—as a symbolic language based on conventions with a high iconic quality, showing scientific, social, technological, and historical connections in pictorial form. In a letter to Stebbing he explained his approach as the "Unification by Logicalisation" and "Unification by Visualisation" in 1939.[12] For Neurath, these were two sides of the same coin and they characterized the scientific agenda of the Unity of science movement and its publication organ, the series "Encyclopaedia of Unified Science", as well as his educational efforts at and through the ISOTYPE Institute, including exhibitions, publications, film for a common public etc. Back in Vienna at the Museum Neurath and his staff had invented the "Method of Pictorial Statistics", on which ISOTYPE is based. They established rules for the systematic transformation of data, first in the context of Red Vienna, basically for the German speaking market. In exile, this was developed further to provide education and information for an international mass audience: No longer primarily the working class and the young generation, but an audience from all social classes was to be addressed. They intended to promote an improvement in the lives of as many people as possible, fostering their ability to analyse and judge the world around them. This was understood to be part of the educational duties and upbringing in democratic societies. During the Second World War, this claim took on a new urgency: the focus was on demonstrating scientific virtues and values in dealing with knowledge and showing what good scientific practice should look like—by practical example. Neurath summarised his approach before his death in a text with the programmatic title "Visual Education: Humanisation versus Popularisation", which was only published posthumously. Here, the close interaction of scientific worldview, socio-political orientation and educational programme becomes clear:

> The transfer of looking at more than one possibility, to be prepared to alter statements, is the principle of the scientific attitude. The social pattern, which permits more than one opinion etc. is the ‚democratic pattern'. Part of education deals with the evolution of one's own judgment, of a ‚scientific attitude', a quality not restricted to scholars only; there are laymen who have it and scientists who do not have it. The transfer of the scientific attitude is not mainly concerned with knowledge but also with the creation of certain sincerity of research and integrity of arguing. (Neurath 1996, 260)

Two aspects should be emphasised in the context of the examination of Stebbing's text on Waddingtons *The Scientific Attitude*, which is kept at the archive in Haarlem together with their correspondence—it can therefore be assumed that Neurath was familiar with Stebbing's comments and criticism. Neurath clearly gives the term

---

[12] Neurath to Susan Stebbing, April 8, 1939 (ONN).

"scientific attitude" a positive connotation here in the sense of an open and reflective attitude towards the acquisition of knowledge as a productive process. The second aspect concerns the status and authority of the knower and the scientific expert, that are both playing a role in the democratic process but are not so much separated because of their specific knowledge. It almost seems as if the quantity and quality of knowledge could be neglected and, instead, the decisive factor seems to be the individual's willingness to develop and accumulate knowledge. At first glance, this understanding also contrasts with Stebbing's view, who describes the roles of experts and laypeople as less fluid. However, the concrete implementation of language criticism and analytical thinking are certainly compatible or overlap. They share the idea that language as a means of communication and basis for intersubjectively communicable statements requires a process of logical clarification and transformation and that clear and analytical thinking can be learned and trained. It is the prerequisite for responsible action in the public sphere—a general "scientific attitude". Moreover, Neurath argues against a "popularisation" of complex scientific content, especially in the service of entertainment and without using a well-thought-out methodology—as does Stebbing. This would only lead to a simplification and falsification of facts. Neurath's ideal is a "humanisation" of knowledge, a didactic build-up from the simple to complex. In order to do this more efficiently and to reach a wider audience with serial publications, the ISOTYPE Institute further developed its methodology and cooperated with various experts and publishers in exile, as they did with the prototypically designed book *Modern Man in the Making* in 1939. Here, Otto and Marie Neurath and their team adapted the method of visualization more closely to the medium of the book. In the preface of his book Neurath described the "picture-text style" especially created for this publication to make the subject dealt with more accessible—a historical-sociological view of the development of modern man, cultures, wars, economy, scientific and medical progress:

> An attempt has been made to evolve for this purpose a special *picture-text-style* which should enable anybody to walk through the modern world that is beginning to appear about us [...] The principle of visualization applied in this book is based on the ISOTYPE method, developed by me together with my collaborators during the last fifteen years. It shows connexions between facts instead of discussing them. Impressive visual aids do not merely act as illustrations or as eye-bait in this book; they are parts of the explanations themselves. The reader may not understand the contents by reading the text only; he must 'read' the pictures as carefully as the text. An international picture language is combined with a world language. (Neurath 1939,7)

The style of the presentation was a combination of a world language, English, and the ISOTYPE picture language: pictorial symbols in combination with other "informative" illustrations such as diagrams, maps, and in later publications also photographs. This was meant to demonstrate a "scientific" way of thinking and train the audience to read and use it, by practical example—supporting people in adopting a modern "scientific attitude" in their daily lives in dealing with increasingly urgent global problems. Neurath and his team intended to create an argumentative form consisting of clear, factual information focussed on the essentials and a systematic symbolic as well as documentary visualization. Especially in an effort to present current events, social issues or data, this was intended to be a more contemporary form of knowledge transfer. For Neurath the democratic system – at least at that time – was the form of society that was inclined towards science, but it needed be actively preserved and defended against its enemies. Especially the planning of change necessary in complex modern democracies required knowledge acquisition and transfer on a global scale, with the help of scientific experts as advisors on the path to self-knowledge, awareness and clear thinking. Decisions still had to be made by the people, to set the goals and make a choice, informed by experts from different fields and disciplines. Here, Neuarth and Stebbing agreed about a clear separation between science and the public or political sphere, between factual statements and value judgements. As Linsbichler and da Cunha recently concluded about Neurath's scientific planning proposals. There is a

> [...] separation between facts, the domain of science, and values, the domain of politics. This separation comes from the realization that science is necessarily a plural enterprise that can support varied points of view. Science would not be able, in Neurath's conception, to offer univocal answers to questions of values, pointing out what would be the best social arrangement. Science's answers cannot even be conceived as absolutely correct. The role of science is only to increase the range of possibilities for dealing with an uncertain future. (da Cunha and Linsbichler 2024, 694)

According to Stebbing in *Thinking to a purpose,* analytical-philosophical thinking and science, a clear and coherent world view would in the long run lead to a strengthening of democracy, to a better understanding of social processes and a stronger participation of citizens. Mastering the difficult process of analyzing and evaluating complex, sometimes divergent opinions and contexts was a duty that no citizen could avoid to preserve the freedom of decision that Stebbing identified as a core element of democratic decision-making:

> My conclusions have been reasonably attained in so far as I have been able to discount my prejudices, to allow for the distorting effects of your prejudices, to collect the relevant evidence and to weigh that evidence in accordance with logical principles. The extent to which I can achieve these aims is the measure of my freedom of mind. To be thus free is as difficult as it is rare. (Stebbing 1939b/1959, 244)

5.4 Disagreements and Common Ground

The implementation of critical-analytic thinking and knowledge transfer for the common reader show many similarities but also some differences between Neurath and Stebbing with regard to the role of science in relation to society and the relationship between scientific experts and laypeople. Pluralist scientific knowledge allows the laymen to reasoning and understanding the different dimensions (social, political, economic, cultural) of a problem and allows for an multidimensional decision process, which might not be the case if only one discipline and scientific approach is included. It also means that it is not only the application of knowledge that guides the decision in the public sphere but the reflection of all possible data—here the difference of the capability and agency of the scientist and citizen comes into play. For Stebbing and Neurath this is connected to paying attention to the use and structure of ordinary language, that is always connected to a certain context and to a certain purpose, to individual interests or groups with a specific tradition, background, culture, and knowledge.

The main differences in the methodology lie in the practical demonstration of the linking of visual and linguistic information to specific topics on Neurath's part, which is intended to encourage critical reflection and clear thinking through demonstration of "good practice". On Stebbing's part, this includes the analysis of examples, which provides scientific authors with information about errors and problems in knowledge transfer and also suggests to laypeople the detailed, critical analysis of different types of texts and the questioning of authorities. Stebbing's interest in Neurath's ideas on visualisation becomes particularly plausible against the background of her own understanding and role in leading the public to logical thinking in the fight against political ideologization. To sum it up: the common denominator between Stebbing and Neurath was and increasingly became the consistent linking of engagement in the public sphere with a more holistic, contextual concept of analytical thinking and practice, mediated by criticism of common language and a pluralistic methodology. And Neurath was very aware of these similarities:

> I clearly realized the tendency within our movement to deal with the actual life when I looked at Stebbing's *Ideas and Illusions*, the preface dated Tintagel April 1941, where the school she and her friends had organized had been evacuated from London. Here she continued her fight against muddled arguing, as started in *Philosophy and the Physicists* and in *Thinking to some Purpose*. But during the war also appeared her *A Modern Elementary Logic* which was intended to be a book for students, some of them in the army, without any guidance from a teacher. I speak of these details, because they clearly show, how persistent scientific life is.[13]

On the philosophical level, however, there were some clear disagreements: Stebbing like other members of the Cambridge School were critical of certain ideas of Logical Empiricism or their implementation. And although Stebbing considered many traditional, philosophical questions as meaningless, she was not ready to accept that the discipline of academic philosophy was just futile, especially as she made various significant contributions to traditional topics. In this sense Stebbing was closer to Moritz Schlick than to Rudolf Carnap or Otto Neurath. On the other hand she criticized certain concepts of Schlick, e.g. the "pursuit of meaning" before the determination of the truth-value of a proposition (Schlick 2013). Another aspect concerned the understanding of scientific analysis on the part of the Vienna Circle and Stebbing's fundamental criticism of their concept of analysis. Since she was interested in refining and specifying the method of analysis in the sense of a clear use of ordinary language, the understanding of analysis on the part of members of the Vienna Circle, did not seem to be sufficiently differentiated. She explained this in the "annual philosophical lecture" of the British Academy in March 1933, titled "Logical Positivism and Analysis". Stebbing compared statements of Schlick and

[13] This posthumously published comment from Neurath is quoted in Chapman (2013), 165.

Carnap to Moore's and Wittgenstein's ideas and found the formers' approach to analysis mistaken in its "methodological solipsism". It would not be the task of philosophy to justify statements or assumptions, but to clarify them. And—so she argued—not only the proposed methodology was the problem, also its objects would have to be determined differently and the way of proceeding rethought:

> We cannot clarify our thoughts by thinking about thinking, nor by thinking about logic. We have to think about what we were thinking about. The philosopher considers a given expression, and analyses it in order to find another expression which says more clearly what the original expression said less clearly. [...] We must first know what facts are the case before we can fruitfully employ analysis for the purpose of clearifying our thoughts about the world. (Stebbing 1933, 36)

Here as also in the context of debates in Cambridge about different types of analysis Stebbing suggested further differentiations: in "postulational" or "logical" analysis, and "directional" analysis as a stepwise form of clarification of the structure of statements, as Stebbing saw it realized by G. E. Moore. This was meant to take into account the context of the use of language, as a purpose-related approach. The correctness or usefulness of an analysis was seen to be relative to the purposes in seeking it. This purpose-relative approach to analysis, as we have heard earlier in the popular book *Thinking to some purpose*, was the one that Stebbing herself came to adopt. For her, the chief task of logic was to make explicit the norms of reasoning. So analysis starts from certain common-sense truths (in the sense of Moore's Non-naturalistic Cognitivism), which then have to be clarified in a process of analyses. In that, Stebbing followed Moore by claiming that knowing the truth of a proposition and the exact meaning of a sentence are two separate issues and philosophy should be primarily concerned with the latter. This can't be further elaborated here,[14] but her contributions were relevant for the ongoing debates before the logical analysis ultimately prevailed. As Michael Beaney (2016) concluded, part of this can be attributed to the growing influence of Logical Positivism. Logical analysis could provide "clearer" formulations of the proposition and elucidate its logical form or remove certain confusions.

In summary, Stebbing and Neurath can be characterised as figures of transition. Both positioned themselves against a one-sided, traditional understanding of science and education marked by social, cultural or religious boundaries. As becomes clear from their correspondence, there was a general willingness to tolerate such differences, in the sense of an alliance against common enemies and threats in times of crisis. Both belonged to a a generation of intellectuals who saw it as their duty to connect knowledge with public impact in service of a future society. It was the conception of the socio-political role of science and the relevance of critical analytical thinking that united Neurath and Stebbing—especially as the ability to weigh evidence and resist hasty conclusions, prejudices and manipulative, emotionalised language was severely challenged at the time. They agreed that the role of the scientific expert was that of a guide on the way to a democratic society that needed the

[14] See also Körber and Tuboly (2024).

clear-thinking, responsible, empowered, and happy individual. Philosophy and the sciences should help to answer the question of a "worthwhile life", as Stebbing expressed in the preface of *Ideals and Illusions* 1941:

> This means that we should have set clearly before our minds what it is we most deeply desire, the attaining of which would bring us inward peace–that is, happiness. In my opinion our most urgent need today is to know clearly what are the things that belong to our happiness. To know this is to begin to formulate a new way of life. (Stebbing 1941, 4)

References

Beaney, Michael. 2016. Susan Stebbing and the Early Reception of Logical Empiricism in Britain. In *Influences on the Aufbau*, ed. Ch. Damböck. Dordrecht: Springer.

Bowler, Peter J. 2009. *Science for All: The Popularization of Science in Early Twentieth-Century Britain*. Chicago: University of Chicago Press.

Burke, Christopher, and Adam Tamas Tuboly. 2024. *Otto Neurath in Britain*. Cambridge: Cambridge University Press.

Carnap, Rudolf. 1963. Intellectual Autobiography. In *The Philosophy of Rudolf Carnap*, ed. P.A. Schilpp, 3–84. LaSalle: Open Court.

Cat, Jordi, and Adam Tamas Tuboly, eds. 2019. *Neurath Reconsidered: New Sources and Perspectives*. Cham: Springer.

Chapman, Siobhan. 2013. *Susan Stebbin and the Language of Common Sense*. Basingstoke: Palgrave Macmillan.

Cunha, Ivan Ferreira da, Linsbichler, Alexander (2024). Science, Imagination and Values in the German Energy Turn: An Example of Neurath's Methodology for Social Technology. Engl. Translation, *Revista Kriterion*, 65 (156), 673-700. https://doi.org/10.1590/0100-512X2023n15603ifc., https://www.researchgate.net/publication/378768837

Doulas, Louis. 2022. Making Sense of Stebbing and Moore on Common Sense. In *Analysis, Common Sense, and Public Philosophy: Themes from the Philosophy of Susan Stebbing*, ed. A. Coliva and L. Doulas. Oxford University Press. (forthcoming).

Janssen-Lauret, Frederique. 2022. *Susan Stebbing*. Cambridge: Cambridge University Press.

Körber, Silke. 2019. Thinking about the 'Common Reader': Otto Neurath, L. Susan Stebbing and the (Modern) Picture-Text Style. *In Cat and Tuboly* 2019: 451–470.

———. 2023. *Die Visualisierung von Wissen im "Jahrhundert des Auges": Otto Neurath, Isotype und Adprint*. Cham: Springer.

Körber, Silke and Adam Tamas Tuboly (2024). Thinking and Muddling for People: L. Susan Stebbing and Otto Neurath. Unpublished manuscript.

Neurath, Otto. 1931/1983. Physicalism: The philosophy of the Viennese Circle. In *Philosophical Papers 1913–1946*, ed. R.S. Cohen and Marie, 52–57. Dordrecht: D. Reidel.

———. 1932/1983. Protocol Statements. In *Philosophical Papers 1913–1946*, ed. Robert S. Cohen and Marie, 91–99. Dordrecht: D. Reidel.

———. 1939. *Modern Man in the Making*. New York: Knopf.

———. 1996. Visual Education: Humanisation versus Popularisation. In *Encyclopedia and Utopia: The Life and Work of Otto Neurath (1882–1945)*, ed. E. Nemeth and F. Stadler, 245–335. Dordrecht: Kluwer.

———. 2013. Erkenntnistheoretische Schriften, 1926–1936. In , ed. Johannes Friedl and Heiner Rutte. Wien: Springer.

Schliesser, Eric (2023). *Stebbing on Clarity*. Unpublished manuscript. https://philarchive.org/archive/SCHSOC-13

Siegetsleitner, Anne 2014. Ethik und Moral im Wiener Kreis. *Zur Geschichte eines engagierten Humanismus*. Wien: Böhlau.

Siegetsleitner, Anne 2015. Susan Stebbing and the Vienna Circle on Moral Philosophy, In: Neuere Aspekte in der Philosophie: aktuelle Projekte von Philosophinnen am Forschungsstandort Österreich, hrsg. von Brigitte Buchhammer, Axia Academic Publishers (= Women Philosophers at work, 1), S. 206–212.

Stadler, Friedrich, and Thomas Uebel, eds. 2012. *Wissenschaftliche Weltauffassung. Der Wiener Kreis*. Wien/New York: Springer.

Stebbing, L. Susan. 1930. *A modern introduction to logic*. London: Methuen & Co. Ltd.

Stebbing, L. Susan. 1933. *Logical Positivism and Analysis. From the proceedings of the British academy*. Vol. XIX. London: Humphrey Milford Amen House, E.C.

——— (1937/1944). *Philosophy and the Physicists*. London: Methuen & Co.

———. 1939a. Ethics and Materialism. *Ethics* 50 (1): 35–44.

———. 1939b. *Thinking to Some Purpose*. London: Penguin Books.

———. 1941. *Ideals and Illusions*. London: Watts & Co.

Verhaegh, S., Petrovich, E., Bös, G. 2024. The American Reception of Logical Empiricism: *A Mention-Based Bibliometric Analysis*, Springer: Cham.

Vlasits, Justin. 2022. Margaret MacDonald's Scientific Common-Sense Philosophy. *British Journal for the History of Philosophy* 30 (2): 267–287.

Vossoughian, Nader. 2003. *The language of the world museum: Otto Neurath, Paul Otlet, Le Corbusier*. Transnational Associations. The Review of the Union of International Associations Spe- cial Issue 1–2:82–93.

Waddington, C.H. 1941. *The Scientific Attitude*. Penguin Books.

West, Peter. 2022. L. Susan Stebbing Philosophy and the Physicists (1937): A re-appraisal. *British Journal for the History of Philosophy* 30 (5): 859–873.

Chapter 6
A Forgotten Chapter in the History of Sociology of Science Philipp Frank's 'Good Nose'

George Reisch and Adam Tamas Tuboly

Abstract This paper reconstructs some conceptual and institutional elements from Philipp Frank's largely forgotten Research Project about the Sociology of Science. The project was organized by Frank at the Institute for the Unity of Science in the mid-1950s, and it involved scholars such as Robert K. Merton, Bernard Barber, Ernest Nagel, and Thomas Kuhn. Due to various circumstances, the group was unsuccessful in developing and shaping the cognitive and professional identity of sociology of science, and this paper highlights some of the possible reasons.

Keywords Philipp Frank · Institute for the Unity of science · Sociology of science · Sociology of knowledge · Thomas Kuhn · Robert Merton · Institutional history

6.1 Introduction

In the last few decades, a number of papers and books have been devoted to the history of philosophy of science. Much less has been written, however, about the history of the intersections of philosophy of science with other disciplines. There are many instances of scientifically oriented philosophers and philosophically minded scientists contributing to neighboring disciplines and fostering each other' causes simply by being interested in something. The sociology of science is one such field—it's true story cannot be told without referencing people who came from the outside and were occasionally rejected because of their allegedly dogmatic philosophy, but who still contributed to the development of the discipline. Their stories are either largely forgotten or entirely unknown in the first place. One important such figure is the physicist-turned-philosopher Philipp Frank (1884–1966).

G. Reisch
MTA Lendület Values and Science Research Group, Budapest, Hungary

A. T. Tuboly (✉)
MTA Lendület Values and Science Research Group, Budapest, Hungary

Department of Behavioural Sciences, Medical School, University of Pécs, Pécs, Hungary

G. Schiemer (ed.), *The Legacy of the Vienna Circle*, Vienna Circle Institute Yearbook 30, https://doi.org/10.1007/978-3-031-80568-4_6

Frank was trained as a physicist in Vienna at the beginning of the twentieth century and led the Theoretical Physics Institute at the German University of Prague (as Einstein's successor) between 1911 and 1938.[1] Due to the growing German threat, Frank arranged a series of talks in the United States from September 1938. As he was crossing the ocean, Germany invaded Czechoslovakia, which obviously complicated his plans. Although he wanted to go back home to Prague, he stayed in the U.S. for the rest of his life. After teaching at Harvard for several years, Frank was able to officially revive Otto Neurath's almost forgotten Institute for the Unity of Science in Cambridge, Massachusetts, with structural support and funding from the Rockefeller Foundation (Reisch and Tuboly 2025).

At the Institute, Frank brought together prominent scholars (sociologists, philosophers, historians) to reflect on the role and nature of the sociology of science. He also established a small group dedicated to the study of certain fundamental problems, in order to advance the field and foster public understanding of science. Eventually, this turned into one of the flagship projects, in the context of post-war general education promoted by Harvard president James Bryant Conant (see Reisch 2017).

In this short paper, we do not intend to secure a place for Frank in the history of sociology of science, and do not aim to paint a picture of smooth cooperation with professional sociologists, or a deep and direct influence on the discipline in the long run. After all, there is a *reason* why Frank is not remembered today; but that reason is not his limited understanding of what sociology of science *is*, or his inability to come up with *ideas* for shaping the field. Rather, the sociology of science project never turned into a full-fledged research program, neither institutionally nor conceptually. Frank was a clumsy organizer; his modus operandi included last-minute invitations, discussions over people's heads, the failure to respect institutional and disciplinary boundaries and territories, and making big promises on which he didn't necessarily deliver. In connection with another planned project, Chadbourne Gilpatric, associate director of the Humanities Division at the Rockefeller Foundation, described Frank's collaboration not only as ineffective (his colleagues, including Carnap, Reichenbach, Feigl or Merton would at times agree), but as "a handicap."[2]

In the following sections, we will provide an overview of how the sociology of science project unfolded. The process was laborious and faced many internal and external problems right from the start. After presenting the main topics, collaborators and results, the paper will close with short case studies detailing how Frank has been neglected. Although we will broach the question of why neither Bernard Barber nor Thomas Kuhn included Frank in their retrospectives on the (institutional) history of twentieth-century American sociology of science, we can only make some preliminary observations here. However, we will explore this issue more

[1] For a short biography of Frank, see Reisch and Tuboly (2021); for all available details, structured in a comprehensive narrative, see Reisch and Tuboly (manuscript). This paper is based on four, much more detailed chapters from the latter manuscript.

[2] Chadbourne Gilpatric to Warren Weaver, November 5, 1952 (RFR, Box 35, Folder 285).

thoroughly in our forthcoming biography of Frank, discussing also more theoretical and scholarly reasons as well (Reisch and Tuboly manuscript).

6.2 Background, Structure and Development of the Project

In this section, we will discuss some basic facts and fundamental issues to get a better sense of the background and structure of the Institute and the project's place within it. While the Institute for the Unity of Science is often dated to 1947, and Frank (1947) indeed published an influential paper that year on its "background and purposes," it took some time for it to be set up. After some initial discussions, which were not without tension, the Rockefeller Foundation provided much-needed financial support, but due to tax issues, the Institute was not able to get the money until 1949, and only launched its official programs in early 1950.

Frank set up various initiatives, from essay contests to get young scholars interested in the Unified Science Movement, to introductory and exploratory workshops and regular general seminars on various topics. But most important for the purpose of this essay were the so-called "subgroups" focusing on specific themes, such as "Cybernetics and Communication" and "Science and Values." They held regular meetings and aimed to explore and develop the field through the involvement of leading scholars. Although "Sociology of Science" was not one of the Institute's official subgroups, it had a similar function. While the cybernetics group gained special significance in the new technological age of the 1950s (see Galison 1998), and the topic of "values" has been particularly important within Anglophone philosophy of science since the 1970s, the Institute had virtually no impact in these areas. Sociology was somewhat different, however.

6.2.1 Dating the Project

The whole project started very laboriously, and it never reached the institutional and organizational level of the mature and complete research project that Frank had envisioned. In fact, for years, his correspondence and reports read as never-ending explanations about the general importance of sociology of science, and as unsolicited excuses about how the Institute's various activities related to this field. Seemingly, Frank took every chance to state that while the Institute had already done a great deal of work in the field, a breakthrough was imminent, if only the financial and institutional support kept coming. At the same time, there were virtually no actual achievements or relevant academic results for years. As more and more people got involved through smaller grants, the research got more and more diverse, and Frank only kept things together due to the necessity of reporting about the Institute's activities.

The project was a long time in the making, marked by disruptions, delays and, most importantly, efforts to recruit an organizing committee and researchers to actually start doing something. In July 1953, Frank reported to Weaver that "since the existence of our Institute, we have had plans to go in for research in the Sociology of Science," and that some preliminary exploratory work had been done with money from the Rockefeller Foundation during the 1952–1953 financial year.[3] It is known from Frank's correspondence that he had planned to work on the sociology of science at least since 1949, occasionally as a "project on the relations between natural and social sciences," with a focus on "The Influence of Social Environment, of Political and Religious Creeds, and Similar Factors upon the Scientific Theories which have been Developed at Certain Periods."[4] In the following 3 years, Frank organized a workshop about the sociology of science with Ernest Nagel and solicited a bibliography of the field (for more on both, see below). In his 1952 correspondence with Nagel, he repeatedly urged him to finally start the project, and to "attack [it] more concretely." And in September 1953, Frank still nagged members like Nagel and Bernard Barber to come up with a clear plan and "final decisions" about the project.[5] For years, one difficulty was to get Robert K. Merton, the biggest name in the field at that time, to actively engage with the Institute. As we shall see below, while Merton was obviously *interested*, he did not put much work into it.

It is difficult to determine when exactly anybody did anything for the project, or how it evolved, but there are a few milestones to be noted. After the 1952–1953 period, Frank again used his Rockefeller Foundation money for the "Research Project in the Sociology of Science," and in July 1954 he informed Weaver of the progress in putting together an organizing committee.[6] In early 1955, Frank noted that the Institute supported various research activities in the sociology of science, and had started work on "quite a few topics which were assigned to a number of research workers." In essence, these small projects concerned potential paradigms and matrices of problems to be pursued in the future. The members of the Institute indeed wanted to get the sociology of science on track by establishing its problems, terminology, and references. But as the deadline for using the Rockefeller Foundation funding was approaching, this time Frank requested not money, but time to be able to make good use of the existing funds. A one-year extension was granted.[7] After it had elapsed, Frank reported in late 1956 that the sociology of science project would from now on be supported by the National Science Foundation.[8] Since little, if any, work was done on the sociology of science within the Institute for the Unity of

[3] "Report of the Institute for the Unity of Science to the Rockefeller Foundation for the Period of July 1952–1953," attached to a letter to Weaver, July 17, 1953 (RFR, Box 35, Folder 285).

[4] Frank to Charles Morris, October 5, 1951 (CMP, Box 4).

[5] Frank to Nagel, February 5, 1952 (CMP, Box 4), April 11, 1952, and September 16, 1953 (AAAS, Box 1101, Folder 33 N).

[6] Frank to Weaver, July 29, 1954 (RFR, Box 35, Folder 285).

[7] Frank to Weaver, March 17, 1955; and Robert S. Morison to Ralph W. Burhoe, April 22, 1955 (RFR, Box 35, Folder 285).

[8] November 13, 1956, Frank to Robert S. Morison (RFR, Box 35, Folder 285).

Science after 1956, it can be said that the project basically ran from late 1953 to 1956 (following an initial exploratory phase), with varying prospects and levels of intensity.

6.2.2 Who Was Involved?

When Frank told Nagel in 1952 that it was high time to finally "attack more concretely" the sociology of science project, he also noted that they should start "with very modest means," giving "to a small number of people an amount of a few hundred dollars each to start to work on some very limited problems." He was about to choose the members of this group. Ideally, they should have some experience in sociology, though this was not a necessary requirement, as Frank indicated: "after all my experience, I do not think that straight sociologists will give us very good suggestions about topics."[9] As most sociologists at that time were not really interested in the nature of science and the (often philosophical) interaction between science and society, Frank's skepticism was not unfounded. One sociologist, however, stood out from the rest in terms of recognition and respect. His name was Robert K. Merton.

The core of the sociology project was a so-called organizing committee, responsible for developing themes and questions, and organizing the research. From the very first moment, Frank was in the lead, followed by Nagel, in his role as a philosopher. Next to his work on the social sciences (Neuber 2022), Nagel also served as a mediator between Frank (who often taught in New York) and Merton. In November 1949, Nagel reported to Frank that Merton was "very much interested" in the project, but had too many commitments and also suffered from health issues.[10] Merton turned out to be a hard nut to crack, as he had various reservations about the membership and structure of the group, but in time Frank took matters into his hands and approached him with a proposal (for details, see Sect. 6.3.1). The idea was, first, that Frank as a *scientist* would formulate the problems and topics to be addressed, before showing them to Nagel *the philosopher*, Merton *the sociologist* and *a historian*. If all agreed, they would then approach a financing body for funding. To give weight to his proposal, Frank noted that "Mr. [James] Conant is very interested in this subject."[11] In this letter, Frank also sent Merton the very first draft of an "outline of a program for research in the sociology of science" (see Sect. 6.3.1).

The possible topics surely caught Merton's eye—many of them were on his own list for future research, as we shall see. Thus, Merton joined Frank's initiative to a certain degree, although he was not involved in any substantial or daily capacity.

---

[9] Frank to Nagel, April 11, 1952 (AAAS, Box 1, Folder 33 N).

[10] Ernest Nagel to Frank, November 12, 1949 (AAAS, Box 1, Folder 33 N); Frank to Ernest Nagel, February 5, 1952 (CMP, Box 4).

[11] Frank to Robert K. Merton, June 1, 1950 (HUA-GHP, Box 62).

The same could be said about Bernard Barber as well, who was Merton's doctoral student, had studied with Talcott Parsons and Pitirim Sorokin, and was professor of sociology at Barnard College for decades. Barber was brought into the project by Merton (to help him out), and they collaborated on several occasions. Nonetheless, Barber was also less committed to Frank's initiative, and as we shall see below (in Sect. 6.5.2), he quite quickly forgot him and his Institute.

There were various people around Frank who established international careers within a few years, after starting out as promising young talents interested in sociological issues. These included Ernst Topitsch, Jeremy Bernstein, Lewis Feuer, and, most importantly, Thomas Kuhn (see also Sect. 6.5.1 below).

6.3 The Idea of Sociology of Science

6.3.1 An Idea Takes Shape

What was the idea behind the project? In July 1953, Frank noted to Weaver that he would soon be occupied with "a field which has been neglected [for] a long time, the field of the Sociology of Science." Indeed, sociology of science was not one of the thriving fields within sociology in the U.S. (partly because of its Marxist leanings, which obviously did not help matters during the Red Scare of the Cold War); Frank thus took some risks and hoped to do some exploratory work, if not by himself, then together with a group of interested experts and adventurous scholars. For Frank, this was a scholarly necessity, given that "the fundamental problems in the Philosophy of Science cannot be solved adequately if we investigate them only from the viewpoint of logical coherence and experimental confirmation."[12] Something more was needed.

To convince others and assemble a group, Frank drafted an "outline of a program for research in the sociology of science" and in 1950 and sent it out to friends and colleagues.[13] Less than two pages long, this proposal is written from the perspective of a scientist baffled by the problem of "how it can happen that scientific theories can be believed not only because they are confirmed by logic and experimental facts, but also by other reasons," such as "social circumstances, political and religious creeds, and similar reasons." The main issue is underdetermination for lack of empirical and logical evidence, as scientific confirmation in the "narrower" and "technical" sense "leaves a wide margin of uncertainty." Frank's idea is that the "turning points" in the history of thought needed to be identified in order to examine the degree to which (Mertonian) "existential factors" had influenced these events.

[12] Frank to Weaver, July 17, 1953 (RFR, Box 35, Folder 285). For details on Frank's sociological thought, see Uebel (2000) and Martin Kusch's chapter in the present volume.

[13] Unless otherwise indicated, all quotes in this section are from the "Outline of a Program for Research in the Sociology of Science," sent to Robert K. Merton, June 16, 1950 (HUA-GHP, Box 62, Folder 4). They are also contained in the Robert K. Merton papers (RKM, Box 26, Folder 13).

His examples are the usual suspects that surface in almost all his writings from this period: Plato's *Republic*, Aristotle, the Middle Ages, the Catholic Church, the modern science of Galileo and Copernicus, and finally, mechanisms of various forms from the eighteenth to the twentieth century, with a special focus on the Soviet Union. He notes that "One should give an objective and complete analysis of the role of scientific theories in the Soviet Union, and particularly an analysis of the social objectives which one wanted to achieve by the regimentation of science." For him, the Hitler regime and spiritualism are also possible subjects of interpretation and analysis, though they are just extreme cases, and a full-fledged project "should analyze to what degree these influences of social compulsion which are so conspicuous under the Soviet regime and under the Nazi regime are existing under all circumstances, except that they are not so noticeable if the government is less authoritarian." However, to arrive at a general picture of the workings of these influences, the best thing to do is to analyze "the public at large," "ruling groups," and all "the masses of such people who have in general little interest in science" while still engaging from time to time with twentieth-century theories.

Frank viewed his points only as a "skeleton" that could be developed into a "plan of research" later on. It was certainly biased, as it represented only the viewpoint of the "scientists" who—as noted above—were "puzzled by the way in which theories have been hailed and attacked, accepted and rejected, on non-scientific grounds."[14] Frank basically invited others to work on these problems, offering his own expertise in science and education and the experience of a long life, during which he had faced extreme political forces and the public reaction to every major scientific event of the century to date. He realized that all these factors and their influence on science were not accidental but went deep into the very heart of the *scientific enterprise*. They thus required a sociologist "to explain that these influences are not just as a scientist often says, a 'nuisance'. The investigation of these influences is just as well a scientific research as the investigation of what one calls usually 'science.'"

At the first major conference of the newly founded Institute, entitled "Current Issues in the Philosophy of the Sciences," held in Boston in April 1950, Frank delivered a paper on "The Logical and the Sociological Aspects of Science" (in the "Science as a Human Enterprise" section). Besides some general recollections of historical examples—again Plato, Aristotle, the Church, Galileo, and Copernicus—the paper is the work of a philosopher rather than a scientist, seemingly a longer elaboration of his "outline." It has two vantage points: underdetermination, and the non-cognitive role science plays in society. Frank starts from the second, showing that already since ancient times, many philosophers, scientists, and political leaders searched for some correspondence between the laws of nature and the laws of society, preferring those scientific theories that supported *a certain, well-defined way of life and behavior*. A theory might be judged for its "truth" or "goodness," based on different criteria, the former being "logical-empirical," and the latter

[14] Frank did much to explain these factors in most of his mature work in the United States. See Frank (1950, 1956a, 1957, 2021).

"sociological-psychological." In Frank's interpretation, Plato—like many leaders—opted for theories that would "be useful for the education of 'good citizens,'" while rejecting those that had damaging social consequences, even if they exhibited certain logical and empirical strengths, such as agreement with observation.

This is rather strange, however, as Frank observes himself: "if a doctrine is confirmed 'by science', how can its social usefulness have an influence on its acceptance?" He argues that "it is not possible to derive a scientific doctrine from observed facts," echoing the points of Poincaré and Duhem. He reiterates certain typical arguments here: If different theories are compatible with the same set of facts, then "logico-empirical" criteria will not be sufficient to determine which one is true. "The authorities, state or church or public opinion, can select from among these doctrines the one which is the most useful for the training of 'good citizens'" (Frank 1951, 19).

But the case for sociology does not end here. Underdetermination of theories "is the most important prerequisite for understanding the role played by sociological factors in the acceptance of scientific doctrines," and there is still work to do to account for certain nuances. Frank thus extends his argument by admitting that while the state, the Church, or any other ruling group (i.e., any form of *authority*) cannot change any observed facts, they can influence whether theories are abandoned or accepted. And this process is the subject of the sociology of science: When science leaves room for decision-making, "existential factors" come into play. Frank calls these moments in the scientific enterprise "stalemates": Situations where logic and empirical considerations cannot help in the decision-making process, so that something else must take their place. He concludes that "All this exploitation of stalemates in science makes a fascinating topic of research in the sociology of science which has been explored very little as yet." (1951, 19, 20, 26).

6.3.2 Research Questions—Part 1

Since no responses from Frank's colleagues to his outline have survived, we do not know how they reacted to it, or how exactly they shaped its development. Two years later, however, another, more detailed and professional proposal was circulated, evidently bearing the imprint of its alleged co-authors, Ernest Nagel, and Robert Merton. Sent to Thomas Kuhn on December 2, 1952, it contained an "outline", a project description in the format of an essay, as well as further questions for research. Although Frank states that the description was written by Nagel, Merton, and himself, all three signatures appear to have been made by Frank, and given his leading role in the project, we can assume that he was the mastermind behind the document.[15]

The essay describing the project is embedded within the "acceptance" problematic that would occupy Frank for years, earning him renewed fame in the late 1950s

---

[15] Frank to Thomas S. Kuhn, December 2, 1952 (TKP-MIT, Box 25, Folder 53).

(Frank 1956a). The question is thus the following: What role do sociological factors play in the acceptance of scientific laws and theories? By "scientific," Frank means the natural sciences, "since the task appears to be simpler in their case than in the social sciences." This is quite a novel move by Frank—since Karl Mannheim's foundational texts in the late 1920s, the sociology of knowledge was generally occupied with the social sciences, and considered physics, mathematics, and related fields as somewhat secured forms of knowledge, not exhibiting any form of acceptable or viable outside influence. Frank obviously broke with this tradition, mentioning that the social sciences should be dealt with at a later point, thereby providing an impetus for exploring new areas.

But how do scientists accept theories? Frank distinguishes between the acceptance of theories because their consequences agree with already established facts (this seems to be the logical, rational, and "commonsensical" scientific way), and acceptance for any other reasons, referred to in the outline as "existential factors," mimicking Merton's problems and vocabulary. This is a very crude and rudimentary description of the project and the field, of course. Nonetheless, it already points to some basic questions that were indeed raised in the essay[16]:

1. Cases of underdetermination, in which different theories have the same empirical results and that the choice of one owes to "actual interference by governments, churches, or other organized bodies."
2. Systematic study of whether "these existential factors are an obstacle to fundamental research" and their possible effects upon "the technological applications of science."
3. Study of how "social utopias" and other "ideal states" may lead to "new forms of organization and regimentation" in science—a topic that suggestively echoes Frank's writings about how physics, biology, and anthropology fared in Nazi Germany.
4. Similarly, but less visibly, cases in which these "very conspicuous" and "extreme forms of social organization" mentioned in (3) may also operate "more subtly even in 'normal' social settings." Here Frank nodded to the popular Freudian methodology for psychoanalytic theory, through which extreme pathological cases are taken to illuminate otherwise ordinary, normal psychological dynamics: "Just as normal psychology may receive much help from the study of abnormality," he wrote, our understanding of "science in more customary societies" may benefit from studying abnormal cases and "what happens under less usual circumstances."
5. Study of "connections" between "types of scientific theories" and "various broad philosophies of life" that may help explain theoretical preferences "shown by those in power for certain types of scientific theory." On the heels of cases (3) and (4), Frank likely means *social or political* power, (distinct from forms of power within research communities that he mentions in cases (7) and (8), below). Nodding again to psychological and ideological dynamics, Frank points out that

[16] For more details and explanations, see Reisch and Tuboly (forthcoming).

these perceived "connections" may have discernible effects whether they are "real or imaginary."

6. One such "connection" is the notion that scientific theories stand as "models for an ideal human society." Where case (3) posits influence flowing from social utopian aspirations into natural science and scientific practice, this case reverses that flow, so that natural scientific theories shape utopian theories and ambitions and consequently underwrite "the attempt to justify social doctrines."
7. The study of how "the social status and social philosophy of natural scientists" affects their scientific work. This extends case number (5), involving theoretical preferences of "those in power," to examine parallel dynamics affecting scientists and scientific activities such as "theoretical constructions." Here Frank also adds "social status" into the mixture of factors for consideration.
8. Having introduced the idea that "theoretical constructions" are influenced by scientists' social philosophies; Frank next opens the door wider to the study of "economic factors on the actual conduct of scientific inquiry." In Frank's one-sentence elaboration, the door swings wider still to include technological, philosophical ideological, pedagogical, and methodological factors (such as "standards of accuracy") as well as broad factors like "intellectual tradition and philosophical preconception, and the like."
9. In his last case, perhaps the most theoretically subtle of all, Frank suggests a reflexive analysis of how these varieties of sociological study would be themselves subject to the sociological influences in question. For "sociological analyses of science also occur within definite social contexts." This idea became known later as the "reflexivity" requirement within sociology of knowledge and rejects Karl Mannheim's ideal of a "free-floating intelligentsia" able to objectively specify social influences without being affected by them.

6.3.3 Research Questions—Part 2

After Frank invited Kuhn to be a member of the committee that oversees the sociology of science project, Kuhn was enthusiastic to accept the invitation. "There is no group with which I should rather participate," he wrote to Frank. Hinting at their previous encounters, Kuhn also noted that "you know of my long-standing interest in the effects of factors transcending the logic of the laboratory upon the formulation and solution of scientific problems". Besides all the formalities, Kuhn hinted at "certain reservations" about the proposed method and topics that could be unimportant for the group, but "might limit" his "usefulness."[17] Kuhn was worried that Frank's outline was too general and too abstract, and thus he wanted to shift the focus from politics, religion, underdetermination, and morality to the commitments of individual scientists to their daily professional groups, such as the American

[17] Thomas S. Kuhn to Frank, December 8, 1952 (AAAS, Box 1, Folder 30 K; also in TKP-MIT, Box 25, Folder 53).

Physical Society. Instead of relying on national communities and other such comprehensive, macro-level groups, Kuhn suspected that

> in twentieth-century western science socially conditioned, implicit, professional "faiths" have assumed many of the roles in the guidance of research and in the determination of the acceptability of scientific theories which religious and metaphysical systems played in the physical sciences of the seventeenth century.

Changing this perspective would reveal how scientists actually considered what the problems that deserve to be attacked and tackled are, what aspects of experiments are relevant, or which criteria a theory shall meet to be valid.

Kuhn and Frank met in person,[18] and in May 1953 Kuhn participated at the meeting of the Sociology of Science Project's organizing committee in New York. A few months later, a new research description, a long list of "possible research topics" was sent to Weaver when Frank reported about the Institute's activities. Illustrating the breadth of the project, it was perhaps designed to be circulated among interested scholars to recruit them for smaller research positions.[19] It has four major parts, (I) "Inception, developments, reception," (II) "The scientist in his professional group," (III) "The scientist in a social system," and (IV) "Special historical problems." Each part features between five and 10 additional sub-questions, with limited guiding information on the possible angles and lines of inquiry (see the table below).

(1) Inception, developments, reception	**(3) The scientist in a social system**
(1.1) Underdetermination, interpretation, historical research, the role of ontology	(3.1) The popular image of the scientist
(1.2) What makes an experiment significant?	(3.2) Variation in the prestige of scientists
(1.3) Resistance of researchers	(3.3) Variation in the prestige of the scientific method
(1.4) The role of analogy	(3.4) Popularization of science (popular and science fiction)
(1.5) Simultaneity of discovery	(3.5) Lay opposition to science
(1.6) The role of youth	(3.6) The relation between ethical and scientific theories
(2) The scientist in his professional group	(3.7) Social and political factors in the reception and interpretation of quantum mechanics
(2.1) Changing standards of science	**(4) Special historical problems**
(2.2) Specialization	(4.1) The interaction between engineering concepts in Carnot's and Joule's physics
(2.3) Research outside the professional area	(4.2) The origin and development of the Ecole Polytechnique
(2.4) The publication process	(4.3) The interaction between science and philosophy regarding the concept of absolute space
(2.5) Research reports	(4.4) The impact of the design of computing machinery on psychological, ethical and epistemological theory since 1600
(2.6) The road from publication to classroom	(4.5) The impact of concepts originating in one scientific field on those of another
(2.7) Comparative analysis of scientific societies	
(2.8) Comparative analysis of scientific funding	
(2.9) Pure science and engineering	

[18] Frank to Nagel, December 9 (1952, AAAS, Box 1, Folder 33 N).

[19] "Possible Research Topics, Sociology of Science," Frank to Weaver, July 17, 1953 (RFR, Box 35, Folder 285).

What does this long list tell us? Many of the topics reflect the interests of Frank and philosophers in general. These include underdetermination, ontology and interpretation; in fact, the whole third section relates mainly to Frank's personal interests and daily work, namely the social reception of scientific theories. Frank contributed to almost all of them in his various papers and books from the 1930s until his death. Other questions were favored more by sociologists (those in part 3 above), or by historians (possibly by Kuhn and Frank; those in part 4), still others by philosophers (like Nagel; those in part 1); certain topics explicitly emerged in later publications as paradigm-shaping lines of inquiry, including the resistance of researchers to discoveries (Barber 1961), and the simultaneity and priority of discoveries (Merton 1957, 1961). Part 2, about professionalism and the societies of the scientists were obviously suggested by Kuhn: they correspond to the points Kuhn made in his unsent letter. This means that for a few months, Kuhn was really close to Frank and the sociology group, and his ideas and suggestions were taken seriously, his topics got included in the proposals that were used to gather a group for the actual research. In the 1950s, Kuhn, Barber, Merton, and even Nagel worked on problems that were listed by Frank, although they never made it explicit that these issues were related in any sense to Frank's project.

As mentioned above, many—and very different—personalities became involved in the project for shorter or longer periods, and their interests (and some of the tangible outputs) resonate well with the above problems. Some topics even seem to be the pet peeves of the people involved, and this potpourri of problems certainly energized some of them. And finally, Frank was never shy about inviting people with even the slightest interest in his problems; he would list their work under the umbrella of the sociology group, helping them not only with money and summer jobs (as was the case with Lewis Feuer and his wife, Ernst Topitsch, Jeremy Bernstein, and many others), but even more importantly, inspiring them to approach scientific issues from a humanistic and social standpoint.[20]

6.4 Results of the Project

What did the Institute achieve exactly, and when did the results materialize? Frank often was much too naïve and optimistic. In March 1955, he notified Weaver that they were behind on their schedule and would need another year to present the results on "a few topics which were assigned to a number of research workers."[21] On January 18, 1956, Frank finally sent something to his new contact at the Rockefeller Foundation. To this "preliminary report," he attached two things: one long

[20] There is obviously no school of Frank, but it would be worthwhile to discuss and reconstruct the history of humanistically oriented philosophers of science, science writers and popularizers, such as Robert R. Cohen, Marx Wartoffsky (from the Boston Colloquium), Jeremy Bernstein, Gerald Holton, and others.

[21] Frank to Weaver, March 17, 1955 (RFR, Box 35, Folder 285).

manuscript about the philosophical interpretations of the theory of relativity," and another on the "hedonistic ethics and the rise of science in the seventeenth century." Frank assured the Foundation that it would "soon receive the continuation of the reports."[22] Most likely, this did not happen; if it did, the reports were lost in the Foundation's archive, which is less probable, given how meticulous its staff were in preserving the smallest invoices, diaries and their notes.

These two manuscripts were not available to the public or to professional colleagues. There were a few exceptions, however. One of the first results of the project—in fact, a preliminary achievement before it was officially launched—was Barber and Merton's "Brief Bibliography for the Sociology of Science" in the second volume of "Contributions to the Analysis and Synthesis of Knowledge," which was edited by Frank and appeared in the *Proceedings of the American Academy of Arts and Sciences*. Merton was approached by Nagel and Frank, and after some initial resistance, he promised to reach out to one of his former students for help—this turned out to be Barber.[23] Merton asked Barber to be the "essential cog in the bibliographical wheel," since this "would fit so nicely with your own manuscript," published soon as *Science and Social Order*. Moreover, since the aim was not to produce a comprehensive bibliography, Merton noted that a general "suggestive working" document, with some annotations and comments, would be sufficient, having "complete authority." A small remuneration (of no more than one hundred dollars) was promised, to be used, according to Merton, "to establish a foundation for the support of research on the sociology of science. If Ford does it, why not we?" Seemingly Merton did not have much confidence in Frank's enterprise, especially compared to the Ford Foundation, that was established on Edsel and Henry Ford's money in 1936 to foster human welfare and funded social research extensively. Merton developed a close relationship with Ford as advisor by the late 1940s, so, he could be in a good position to know that by mid-1950, the Foundation has decided to launch a new Behavioral Sciences Program that would run, in fact, until 1957.[24]

Barber was happy to accept the invitation to collaborate—in fact, it was an invitation to *work*, since Merton made it clear well in advance that he only had time to fill in some gaps. Utilizing his class materials, Barber jumped in, drafting the introductory essay, and updating his reading list to create a bibliography.[25] When the bibliography (Barber and Merton 1952) appeared (the introductory essay was in many ways a mixture of Barber's book and Merton's previously published work *Social Theory and Social Structure*), it was everything but critical and annotated. Noting in each of their letters to each other that they did not have much time, Barber and Merton viewed this as a basic task to deliver as quickly as possible. Essentially, they just listed, on more than ten pages, the most relevant literature under specific headings, such as "general theory in the sociology of science," "cultural structure of

22 Frank to Robert S. Morrison, January 18, 1956 (RFR, Box 35, Folder 285).

23 Nagel to Frank, November 12, 1949 (AAAS, Box 1101, Folder 33 N).

24 Robert K. Merton to Bernard Barber, November 21, 1950 (RKM, Box 6, Folder 12).

25 Bernard Barber to Robert K. Merton, November 30, 1950 (RKM, Box 6, Folder 12).

science," "the social organization of science" (recruitment and social composition, organizations and the organization of actual work by the university, industry, government and foundations), "the relations of science and other social institutions" (like religion and economic and political institutions, with emphasis on the Nazi and Soviet regimes), "social and psychological components of scientific work: invention and discovery," and "the social consequences of science."

The latest entry on the list dates to 1951, and the first to 1896 (an article by Mach on accidents in scientific discovery), and it focuses almost exclusively on writings in English, mainly American books and articles, plus some translations of the inevitable classics. Besides the usual suspects, including Talcott Parsons, Max Weber, Michael Polanyi, Thornstein Veblen and Merton himself (but interestingly not Mannheim), it includes even such works as Sarton's on history of science, Bronislaw Malinowski's on anthropology, an autobiographical text by Percy Bridgman, and Eugene Wigner's volume on human values. For whatever reason, Frank did not appear on the list, despite his work, like his 1951 paper on the "logical and sociological aspect," clearly matched the general intent of the bibliography.

In fact, Frank's paper and orientation resonated well with Barber and Merton's aims, though Frank's were more expansive, especially his forays into history and philosophy. In their introduction, which seemingly serves as a replacement for annotations, Barber and Merton wrote that the subject matter of sociology of science is "*the dynamic interdependence* between science and society, as an ongoing social activity giving rise to empirically confirmed rational knowledge, and the environing social structure." This dynamic interdependence goes from science to society, which has often been noted ("Hiroshima only verified what everyone knew. Science has social consequences"), while the "impact of society upon science" has been neglected (Barber and Merton 1952, 140). Frank obviously called attention to the latter.

Barber and Merton are careful to outline their starting points: They acknowledge that the social structure influences "the rate of development, the foci of interest and, perhaps, the very content of science." The last point was, and still is, much disputed, despite decades of research clearly showing how social issues in a broad sense are reflected in the content of scientific knowledge. People worried as much about the autonomy of science in the 1950s as they do today: The authors call the idea that social influence would jeopardize that autonomy a "mistaken belief." They also make a more general point in relation to politics:

> There may be something here of the sentiment that science remains the more pure and unsullied if it is implicitly conceived as developing in a social vacuum. Just as the word "politics" now carries for many the connotation of base corruption, so the phrase "social contexts of science" may connote for some physical scientists the intrusion of concerns alien to science proper. (Barber and Merton 1952, 141)

Another possibility would be that if one admits that science, and especially scientists, are influenced by non-scientific factors, then the desire to advance knowledge would be replaced by "aggrandizement of the self." But those who advocate this opinion conflate "the level of motivational analysis for the level of institutional analysis." In fact, several different motivations may be equally at work in a given institutional framework (people may pursue pure research for money, curiosity, prestige or power), and the task of sociology of science is indeed to identify all these factors and their impact. Naming and categorizing the social influences *on* scientists is not the same as to "arraign" them (Barber and Merton 1952, 141).

Yet again, Barber and Merton note the typical case of Nazi Germany, where society and politics had a "dramatic impact upon the nature, quality, and direction" of science. Nonetheless, they do not agree that this an "exceptional and pathological case" without any further implications; rather, Barber and Merton consider the cases of Nazi Germany and the Soviet Union to be extremes that shed a light on more general tendencies. Just like Frank, they also believe that no society can avoid exerting an influence on science; while the Nazis and the Soviets are extremes, they are located on a continuum with liberals, democrats, or socialists in the U.S.

What had previously been seen as "Marxist figment" became a common experience of scientists and citizens during and after World War II, though not much actual research was done on the subject. Concluding their introduction, Barber and Merton (1952, 141) mention "ample institutional evidence" for the failure to scientifically examine the society-science relation: "nowhere among the universities of this country is there an Institute for Research on the Social Relations of Science." However, this was just about to change, and they knew it. So why did they fail to mention or to reflect on Frank's efforts in this direction? Perhaps because they either did not trust that he would succeed, or because they felt that his nascent group would be ineffective on account of its method and topics. In hindsight, they did not make a great mistake with this omission; nonetheless, no one knows what would have happened if Barber and Merton had lent their full support to Frank and his cause.

A year before the bibliography was published, Frank organized a smaller workshop, one of the first events of the recently established Institute for the Unity of Science. Held at the American Academy of Arts and Sciences in Boston from May 26 to 27, 1951, it was dedicated to "Logic and Sociology of Science." While the first day was devoted to logic and Quine's recent "Two Dogmas of Empiricism," the second centered on "The Sociological Aspect of Science." The idea was not to organize any formal lectures, but to encourage the audience to actively participate in the discussion. Thus, after a short introductory speech, the panel members would reflect on the prepared papers for 10 min.[26] After Frank had suggested organizing the meeting around the sociology of science, Nagel was skeptical as to who could lead the discussion, given Merton's general unavailability and their struggles in organizing the research group during the previous year. He suggested the sociologist C. Wright Mills from Columbia University, the economist Paul Samuelson from the Massachusetts Institute of Technology (MIT), Karl Deutsch, the philosopher George Boas from Johns Hopkins University, the psychologist Paul Meehl from the University of Minnesota, the historian Edward Rosen from City College, and the philosopher and sociologist Edward Strong from the University of California,

[26] Philipp Frank to Bernard Barber, May 22, 1951 (BBP, Box 26, "Unity of Science, 1951 May").

Berkeley.[27] As is immediately apparent, this list is rather heterogeneous, confirming the concerns of many professional sociologists that sociology of science was done by scholars working on the periphery, without many professional credentials in the discipline.

Interestingly, except for Deutsch, none of the individuals proposed by Nagel made it to the conference. The introductory speeches were given by Deutsch and by Barber, whom Frank had invited just a week before the workshop (not leaving him much time to come up with a well-thought-out, substantive speech). Panel members included I.B. Cohen, P.W. Bridgman, Charles Morris, Gerald Holton, and the Jesuit philosopher of science Joseph T. Clark (who was Frank's student previously); Nagel was also absent. Although twice as many people participated on the first day as on the second, it still attracted 33 attendees, among them Robert R. Brockhurst, Arnold S. Kaufman, Albert Hofstadter, Yehoshua Bar-Hillel, William Craig, Giorgio de Santillana, and Thomas Kuhn.[28]

In his "introductory remarks," less than a page long, Barber notes that the meeting—based on the idea that science is just as much a social behavior as any other human activity—"represents an important formal recognition of the need for a sociology of science." Thus, the aim of the meeting should not be to debate the legitimacy of the field, or to provide arguments for the validity of its viewpoint, but rather—more positively—"to discuss analytical tools for the useful understanding of social uniformities and social variations in scientific behavior." The question is whether the meeting achieved anything of the sort, given that the panel consisted not of professional sociologists, but of scholars with a strong interest in social perspectives. Barber, as an optimistic professional, emphasizes that sociology of science is a specialization of "systematic, generalized sociological theory" and could therefore "never be any better than sociology of a whole." For his part, Frank was surely not interested in sociology of science as a subdiscipline of general sociology, but rather as a perspective to describe and understand the multifaceted character of science.

Before ending his notes, Barber claims (relying on a condensed chapter from his soon-to-be-published book) that he would show how social values and institutional structures—as sociological categories—"apply to scientific uniformity and variation in 'liberal' and 'authoritarian' societies in the modern world," with the aim of providing "some basics for discussion."[29] Barber surely experienced a strange mixture of responses about his detailed sociological work that might have been hard to follow for the physicists. We do not have any evidence about the success of the

[27] Ernest Nagel to Frank, March 2, 1951 (YNP).

[28] Attendee list of the "Logic and Sociology of Science Conference" (AAAS, Box 1101, Folder 8).

[29] Bernard Barber, "Introductory remarks" (BBP, Box 26, "Unity of Science, 1951 May"). Frank's edited volume, *The Validation of Scientific Theories* (1956b) could also be seen as an outcome of the sociology project, though it was done under the general auspices of the Institute and will be discussed in detail in Reisch and Tuboly (forthcoming). See also Reisch and Tuboly (2021) for a discussion of Frank's last book manuscript (Frank 2021), which he had originally envisioned to be part of his sociology project but never finished.

sociology workshop, or about the direction of the actual discussions that took place among Barber, Frank, Deutsch, and the audience.

6.5 Frank's Fate Within Sociology of Science

There may be countless reasons why Frank fell out of the canon (given that he was part of it for a short time), or why he has been forgotten in the institutional history of sociology of science. Some of these reasons are of a philosophical, conceptual, or methodological nature. After all, Frank did not pursue any of his projects according to the standards of the field: He never conducted any empirical research, and his historical and personal anecdotes were not scientific enough for Merton and Barber, who considered those stories to be signs and expressions of amateurish outsiders. But this is not the line we would like to pursue here.[30]

We will close this paper with two examples that clearly show that the neglect of Frank started already during his lifetime, and continued for decades, even among many of his close colleagues whom he had helped. It cannot be excluded that some people ignored him intentionally, but in other cases it was simply an unconscious move that Frank's name was forgotten, or rather did not come up when it should have.

6.5.1 Kuhn's Neglect of Frank

So far, we have not discussed Frank's relation to Kuhn, which deserves a study of its own (see Reisch and Tuboly forthcoming). Suffice it to say that Frank and Kuhn were close for many years, and that Kuhn knew a great deal about Frank's group, having been involved in it to some extent; he even asked for Frank's opinion on certain issues that emerged when he wrote his *Structure of Scientific Revolutions* (see Reisch and Tuboly 2021, 50–52, Bentley (2024)). It would thus be wrong to say that Kuhn had no personal relationship with Frank, or that he simply forgot about him. In fact, even in the mid-1960s, when speaking about the royalties from *Structure* (which was published in the *International Encyclopedia of Unified Science*, then part of Frank's Institute), he referred to Frank's difficult personal situation. After recalling that Frank was allegedly living "penuriously" from the royalties, Kuhn wrote that he had "no intention of upsetting the apple cart, for I love [Frank] dearly and sympathize with his plight."[31]

Nonetheless, there is a telling story about Kuhn's relation to Frank, especially in the context of sociology of science. In early December 1959, Pendleton Herring,

---

[30] For more details about these methodological divergences and differences, see Tuboly (2017) and Reisch and Tuboly (forthcoming).

[31] Thomas S. Kuhn to Maurice English, January 21, 1965 (TKP-MIT).

president of the Social Science Research Council (SSRC), approached Kuhn for suggestions about the "further development of a sociological approach to the history of science," which Kuhn had mentioned at a conference a few weeks earlier.[32] The conference was the SSRC's major event on the "History of Quantification," organized together with the National Research Council Committee on the History of Science. The theme of the conference was the meaning of quantification, both as a tool in the "perpetual dialogue between nature and the scholar," and as a method for achieving higher levels of abstraction. To this end, it included lectures about the nature and role of quantification in the sciences in general (Harry Woolf), in medieval physics (A.C. Crombie), in modern physics (Kuhn), in chemistry (Henry Guerlac), in the medical sciences (Richard H. Shryock), in psychology (Edwin G. Boring), in economics (Joseph J. Spengler), in biology (R.W. Gerard), and finally in sociology (Paul F. Lazarsfeld). But these are just the general, invited papers, published 2 years later in a separate volume of *Isis* (see *Isis*, Volume 52, Number 2, June 1961). As stated in the introduction to this special issue (Woolf 1961, 134), the conference was attended "by some thirty scholars representing eleven academic disciplines," among them Philipp Frank of Harvard.

Nonetheless, in his reply to Herring after the conference, Kuhn makes no mention of Frank whatsoever. Instead, he talks about his own "suggestions about ways of promoting the sociology of science." Even though the field had already matured during the 1950s, thanks to foundational work done by Merton, Barber and Frank, Kuhn notes only how fruitful the sociology of science "could become in a very short time, given a few people working in it." The potential is there, especially for a "national policy," by doing away with "stereotypes" and replacing them "with more viable concepts." But according to Kuhn, this is only a promise, not yet a reality—obviously, his own forthcoming book would be a major step in that direction, deconstructing many stereotypes and misleading ideals.

Herring also solicited Kuhn's opinion on planned actions and how to turn the subject into a respected field. In his response, Kuhn says that he does not "understand why there has not already been far more activity in it." He suspects that this is because the people "with the requisite scientific background" who had emerged within the sociology of science in the previous decade unfortunately always turned "to problems in statistical analysis or to discussion of the nature of the methods capable of making sociology more 'scientific.'" What these people were doing was to apply "science to sociology" rather than "sociology to science," which is why they first needed to be persuaded to do the opposite. It is apparent that Frank was not on Kuhn's mind at this point.

According to Kuhn, whenever a few people with the required background and the right intentions get together, they should produce "publicized evidence" of interest, "productive work within the nation's best sociology departments" that would also attract new minds. Though this would happen "inevitably," he believes, "some

[32] Pendleton Herring to Thomas S. Kuhn, December 7, 1959 (TKP-MIT, MC240, Box 23: Lecture/Meetings: Berkeley Social Science Colloquium, 1956.)

strategically applied pressure might accelerate" the process. The best strategy would be to assemble "sort of a summer study group" for people in sociology, history of science and economic history. As an "intermediate step," these scholars should focus on "Outstanding Problems in the Sociology of Science and the Resources Available for their Study," before big projects and institutes could be launched. However, Kuhn reserves any further details for a personal discussion, if would be sufficiently committed. As a start, Kuhn suggests a meeting including Robert K. Merton, Bernard Barber, and himself.[33] Kuhn respected Merton in a sense, and they corresponded about their work, which Kuhn views as "just the sort of work that could make the sociology of science a very worth while field indeed." He only sees one problem with this: "how are we to get some bright people working on it?"[34]

A study group with five to ten members, featuring Merton, Barber, and Kuhn, to address "outstanding problems in the sociology of science" was almost literally what Frank had envisioned and organized in the 1950s. One cannot really say that he accomplished it, but he did not focus on "science on sociology" (Kuhn's charge against others in the field), but the other way around. Nonetheless, it either never crossed Kuhn's mind to mention Frank's efforts (although in an unsent letter drafted before joining Frank's committee, he engaged seriously and across multiple pages with Frank's ideas about the level and scope of sociology of science) or the study group Frank had organized a few years earlier (which Kuhn had helped to shape here and there), or he deliberately neglected Frank because of his failure to live up to the project's ambition. Be that as it may, Kuhn certainly didn't consider Frank's research project serious and substantial enough to commit to it in any way.

6.5.2 The Selective Memory of Sociologists

In the late 1980s, Barber was about to publish a Chinese translation of his *Science and the Social Order*, with an updated introduction, summarizing the historical development of sociology of science as a discipline. In fact, this introduction was published in *Science & Technology Studies* as "The Emergence and Maturation of the Sociology of Science" before the translation appeared. In it, Barber tries to explain the origins and context of his book and how the field emerged in the early 1950s, when, according to him, most people were primarily interested in "practical policy, not scholarship for its own sake."

When he read this, Merton was furious because he thought that the paper was misleading, did not do justice to various people, especially to himself, and failed to properly represent the history of the field. In particular, he took issue with the following passage:

---

[33] Thomas S. Kuhn to Pendleton Herring, December 21, 1959 (TKP-MIT, MC240, Box 23: Lecture/Meetings: Berkeley Social Science Colloquium, 1956).

[34] Thomas S. Kuhn to Robert K. Merton, September 16, 1958 (TKP-MIT, MC240, Box 22: Correspondence: Merton, Robert K.).

> As an undergraduate student of Merton's in the late 1930s, and as his associate and close friend thereafter on a continuing basis during the 1940s and 1950s, I heard nothing of the sociology of science as a specialty … [U]ntil the 1960s the sociology of science did not exist nor was even conceived of as a specialty in sociology… (Barber 1987, 130–132)

Merton then typed up eight long pages filled with recollections, quotations from published materials and from his correspondence with Barber to rebut the latter's erroneous recollections.[35] "What is one to make of that statement," asks Merton, "and others of like kind which turn up in this piece?" Merton has a few things in mind: First and foremost, he quotes extensively from his own publications before and during the 1950s. He tries to point out that in his work, he was calling attention to the sets of problems within sociology of science (in *Social Theory and Social Structure*, for example, which had a whole section on "Studies in the Sociology of Science"). He thus demonstrates that "evidently, by the late 1940s, one of us had been thinking of 'the sociology of science as a specialty.'" Consequently, Merton argues that "sociology of science as a specialty" had occurred to him long before the 1950s (he points to some of his papers from the 1930s), and that, by implication, all of his students must have been aware of it as well.

But that is not all. Merton calls Barber's attention to the fact that if anyone were to read their archives in the future, they would find even more evidence for the misleading nature of Barber's statements about the field. He quotes from one of his letters to Barber from 1950, in which he invited him to work on the bibliography for the Institute for the Unity of Science. In this letter, Merton had noted two things. First, a recognition of the "cognitive identity of the field and a sense of the need for its institutionalization." Recall that Merton had proposed to spend their fee on establishing a "foundation for the support of research on the sociology of science."[36] As such, Merton claims, there was evidently a conception of sociology of science, especially since there was a need to represent it with a bibliography. Secondly, it wasn't just any bibliography, but well-selected material to be published in a well-chosen publication. Here are Merton's comments:

> As for the auspices of this proposed bibliography, you will recall that Tom Kuhn's incomparably consequential STRUCTURE OF SCIENTIFIC REVOLUTIONS was initially prepared under the auspices of the INTERNATIONAL ENCYCLOPEDIA OF UNIFIED SCIENCE, appearing a dozen years later as Volume II, No. 2 of the ENCYCLOPEDIA and, more notably, as a free-standing volume read the world over.[37]

The implication here seems to be that Frank's ventures, like the Encyclopedia, were serious and respectful enough at the time. It was where Kuhn published his most important book, after all. And if it was good enough for Kuhn, they, that is Merton and Barber, should have been just happy to engage with Frank and the Institute. This

---

[35] "Notes on Early RKM References to the Sociology of Science," Merton to Barber, April 6, 1988 (RKM, Box 6, Folder 12).

[36] Barber to Merton, November 21, 1950 (RKM, Box 6, Folder 12).

[37] "Notes on Early RKM References to the Sociology of Science," Merton to Barber, April 6, 1988 (RKM, Box 6, Folder 12).

is some very interesting *retrospective reasoning*—while at the time, Merton had noted repeatedly to Barber that the bibliography was intended for the unity-of-science movement and its followers, rather than being aimed at sociologists of science, neither of them had ever emphasized how important for them and the field of sociology of science Frank's initiatives were.

Merton recalls that in his original preface to Barber's book, he had made a "swift allusion to Philipp Frank and the workgroup of the America Academy of Arts & Sciences (with which we were affiliated)." But more important for him is the following rhetorical question: "what had I been doing, in the 1930s, 40s and 50s, in publishing a monograph and an array of papers focused on a variety of stated problems in the sociology of science?"[38]

In his prompt reply, Barber tries to restrict the whole issue to one sentence—Merton repeatedly emphasized that there were at least two misleading passages and not just one sentence—and to some general "concerns", downplaying the importance of Merton's points. Merton recognized Barber's strategy and thus highlighted all mentions of the word "concern" in Barber's letter to show how desperately Barber tried to dismiss his actual worries. Barber explains that his neglect of the field's early history was about "showing the absence of either an informally or formally organized professional group working in the sociology of science."[39] Yet again, there is no mention of Frank or his group—even though Merton warned Barber about their affiliation to Frank's sociology group and Institute back then. A month later, Barber made a "correction" and sent it to Merton, who again annotated it heavily, as Barber had restricted his correction remarks—yet again or still—only to "one sentence" in the original introduction, thereby again downplaying Merton's points. But now Barber emphasized more concretely the absence of a "professional identity" of the sociology of science, given that the field lacked—even until the 1970s—important elements, like "journals, professional associations, regular formal professional meetings, dense networks of informal interactions and footnote citations among the professional specialists, honorific prizes, and specialized university courses." This correction also appeared in print (Barber 1988).[40] It shows that Barber did not really take Frank seriously; even after Merton brought up Frank's name, Barber did not mention him, neither in the correction nor in his letters to Merton.

[38] "Notes on Early RKM References to the Sociology of Science," Merton to Barber, April 6, 1988 (RKM, Box 6, Folder 12).

[39] Barber to Merton, April 11, 1988 (RKM, Box 6, Folder 12).

[40] Barber to Merton, May 26, 1988 (RKM, Box 6, Folder 12).

6.6 Conclusion

Frank's project on the sociology of science helped a few people, either by providing them with temporary affiliation or small grants, but perhaps even more importantly, food for thought that in some cases lasted for decades. He did much to foster the "cognitive identity" of this recently established field, though its "professional identity" was not even in question yet. Especially after Kuhn entered the scene, followed by many other scholars in the 1960s, Frank's wide-ranging, but loosely organized initiative evidently lost its appeal when scientific communities moved from a cognitive to a professional identity.

However, we should not forget all his efforts and achievements, which certainly had their place at the time. In fact, Gerald Holton did just that in his correspondence with Merton. After the latter showed him Frank's invitation letters from the 1950s, Holton highlighted (referring to "our dear Philipp Frank") "Frank's usually well-hidden but immensely creative way to mobilize someone for the 'Institute', get [the National Science] Foundation involved. But above all it shows what a 'good nose' he had (Einstein called him "ein guter Spürhund") for discovering where a lively discipline exists to which he can attach the 'Movement' while helping the discipline to grow. How we miss him!"[41]

6.7 Archival Sources

AAAS	Record Group XXI: Academy Projects and Programs. Institute for the Unity of Science (1947–1959). Archives, American Academy of Arts and Sciences, Cambridge, Massachusetts.
BBP	Bernard Barber Papers, 1938–1988, Rare Book and Manuscript Library, Columbia University Library.
HUA-GHP	Harvard University Archives, Papers of Gerald Holton, 1919–2015 and undated (HUM 132).
CMP	Charles Morris papers, Institute for American Thought, Indiana University—Purdue University, Indianapolis.
RFR	Rockefeller Foundation records, projects, SG 1.1, series 100, international—series 257 Virgin Islands, U.S.A. (FA386a), international—general (no program), subseries 100.Gen.
RKM	Robert K. Merton Papers, 1928–2003, bulk 1943–2001. Rare Book and Manuscript Library, Columbia University Library.
TKP-MIT	Thomas S. Kuhn papers, MIT Institute and archives.
YNP	Yvonne Nagel papers. Personal collection.

[41] Gerald Holton to Robert K. Merton, April 22, 1997 (HUA-GHP, Box 62, Folder 4).

References

Barber, Bernard. 1961. Resistance by Scientists to Scientific Discovery. *Science* 134 (3479): 596–602.

———. 1987. The Emergence and Maturation of the Sociology of Science. *Science & Technology Studies* 5 (3/4): 129–133.

———. 1988. Correction. *Science, Technology, & Human Values* 13 (1–2): 215.

Barber, Bernard, and Robert K. Merton. 1952. Brief Bibliography for the Sociology of Science. *Proceedings of the American Academy of Arts and Sciences* 80 (2): 140–154.

Bentley, Jospeh. 2024. Positivist or Post-Positivist Philosophy of Science? The Left Vienna Circle and Thomas Kuhn. *Studies in History and Philosophy of Science* 107: 107–117. https://doi.org/10.1016/j.shpsa.2024.08.003.

Frank, Philipp. 1947. The Institute for the Unity of Science: Its Background and Its Purpose. *Synthese* 6 (3/4): 160–167.

———. 1950. *Relativity – A Richer Truth*. Boston: Beacon Press.

———. 1951. The Logical and Sociological Aspect of Science. *Proceedings of the American Academy of Arts and Sciences* 80 (1): 16–30.

———. 1956a. The Variety of Reasons for the Acceptance of Scientific Theories. *The Scientific Monthly* 79: 139–145.

———, ed. 1956b. *The Validation of Scientific Theories*. Boston: Beacon Press.

———. 1957. *Philosophy of Science: The Link between Science and the Humanities*. Englewood Cliffs, NJ: Prentice Hall.

———. 2021. *The Humanistic Background of Science*. Ed. by George Reisch and Adam Tamas Tuboly. Albany: SUNY Press.

Galison, Peter. 1998. The Americanization of Unity. *Daedalus* 127 (1): 45–71.

Merton, Robert K. 1957. Priorities in Scientific Discovery: A Chapter in the Sociology of Science. *American Sociological Review* 22 (6): 635–659.

———. 1961. Singletons and Multiples in Scientific Discovery: A Chapter in the Sociology of Science. *Proceedings of the American Philosophical Society* 105 (5): 470–486.

Neuber, Matthias. 2022. Nagel on the Methodology of the Social Sciences. In *Ernest Nagel: Philosophy of Science and Fight for Clarity*, ed. Matthias Neuber and Adam Tamas Tuboly, 215–232. Cham: Springer.

Reisch, George. 2017. Pragmatic Engagements: Philipp Frank and James Bryant Conant on Science, Education, and Democracy. *Studies in East European Thought* 69: 227–144.

Reisch, George, and Adam Tamas Tuboly. 2021. Philipp Frank: A Crusader for Scientific Philosophy. In *The Humanistic Background of Science*, 1–67.

———. 2025. Reviving the Unity of Science Movement: Philipp Frank's Journey to Harvard. In *American Philosophy and the Intellectual Migration: Pragmatism, Logical Empiricism, Phenomenology, Critical Theory*, ed. Sander Verhaegh. Berlin: DeGruyter, 169–190.

Reisch, George, and Adam Tamas Tuboly (manuscript). *Humanistic Philosophy of Science: Philipp Frank's Life Between Science and the Humanities*. Unpublished manuscripts.

Tuboly, Adam Tamas. 2017. Philipp Frank's Decline and the Crisis of Logical Empiricism. *Studies in East European Thought* 69: 257–276.

Uebel, Thomas. 2000. Logical Empiricism and Sociology of Knowledge: The case of Neurath and Frank. *Philosophy of Science* 67: 138–150.

Woolf, Harry. 1961. The Conference on the History of Quantification in the Sciences. *Isis* 52 (2): 133–134.

Chapter 7
The American Reception of Logical Empiricism: A Mention-Based Bibliometric Analysis

Sander Verhaegh, Eugenio Petrovich, and Gregor Bös

Abstract This paper supplements existing work on the development of logical empiricism with new quantitative data concerning the movement's reception in the United States. Using EDHIPHY ("Enriched Data for the History of Philosophy"), a relational database specifically designed for bibliometric research on the development of twentieth-century philosophy, we trace the impact of logical empiricism on American philosophy before and after the migration. Specifically, we make use of EDHIPHY's mention index, a catalogue of 1,095,765 mention links extracted from 22,977 full-text articles published in 12 Anglophone philosophy journals between 1890 and 1979 and conduct a comparative analysis of these mentions over time. We (1) explore the relative impact of 11 prominent representatives of logical empiricism, (2) compare the movement's reception with a number of contemporaneous philosophical schools, (3) map its evolving place in the American philosophical landscape through a series of co-mention analyses, and (4) dissect logical empiricism's reception by exploring the role of various journals and universities.

Keywords Logical empiricism · Digital humanities · Bibliometric analysis · EDHIPHY · Analytic philosophy · American philosophy · Intellectual migration

S. Verhaegh (✉) · G. Bös
TiLPS, Tilburg University, Tilburg, The Netherlands
e-mail: A.A.Verhaegh@tilburguniversity.edu; G.E.Boes@tilburguniversity.edu

E. Petrovich
TiLPS, Tilburg University, Tilburg, The Netherlands

University of Turin, Turin, Italy
e-mail: E.Petrovich@tilburguniversity.edu

G. Schiemer (ed.), *The Legacy of the Vienna Circle*, Vienna Circle Institute Yearbook 30, https://doi.org/10.1007/978-3-031-80568-4_7

7.1 Introduction

Logical empiricism has its origins in Europe but was most successful in the United States. It emerged from the activities of several groups of scientific philosophers in, among others, Berlin, Prague, and Vienna and became one of North America's leading intellectual currents after some of its foremost representatives—e.g. Rudolf Carnap, Herbert Feigl, Philipp Frank, Carl Gustav Hempel, and Hans Reichenbach—sought refuge across the Atlantic in the years before the Second World War. The logical empiricists and their American followers helped institutionalize philosophy of science as a stable philosophical (sub)discipline and played an important role in the development of analytic philosophy, still the dominant approach to philosophy in the United States today (e.g. Giere 1996; Stadler 2007; Richardson 2017; Dewulf 2021; Verhaegh forthcoming).

Logical empiricism's successes raise a number of questions concerning the development of mid-twentieth century American philosophy. In recent years, historians have tried to answer these through detailed studies of the movement's reception history (e.g. Hardcastle et al. 2003; Reisch 2005). They have traced logical empiricism's first encounters with American philosophers in the years before the migration (e.g. Uebel 2015, 2016; Verhaegh 2023, 2024a). They have uncovered the empiricists' activities and the responses of their new American colleagues (e.g. Jewett 2011; Limbeck-Lilienau 2012; Misak 2013; Verhaegh 2020a, b). And they have developed new, more critical perspectives on the movement's reception, noting that important elements of its original program were lost in translation (e.g. Howard 2003; Reisch 2005; Uebel and Limbeck-Lilienau 2022). While Carnap is one of the most-discussed philosophers in the analytic tradition, much of his program appears to have been misunderstood by his American students and critics (Richardson 1998; Friedman 1999). Other currents within the logical empiricist movement, most notably the perspectives of Frank and Neurath, were altogether ignored by the analytic mainstream (Tuboly 2020; Reisch and Tuboly ms.).

As yet, most work on logical empiricism's reception is based on literature study and archival research. These methods are well suited to conduct relatively fine-grained analyses but are less useful if one seeks to map logical empiricism's place in American philosophy more broadly. While close reading and archival research allow historians to develop detailed reconstructions of the activities of (small networks of) philosophers and the evolution of specific concepts, theories, and debates, they have limitations when one seeks to investigate bigger-picture claims. Historians can only analyze relatively small sets of publications and documents and are thus forced to examine this period through a magnifying glass, making it difficult to establish to what degree their sources are representative of the broader philosophical conversation (cf. Braat et al. 2020, 2).

This paper aims to supplement existing work on the development of logical empiricism with new quantitative data concerning the movement's reception in the United States. Using EDHIPHY ("Enriched Data for the History of Philosophy"), a relational database specifically designed for bibliometric research on the development of twentieth-century philosophy (Petrovich et al. 2024), we trace the impact of logical empiricism on American philosophy before and after the migration. Specifically, we make use of EDHIPHY's mention index, a catalogue of 1,095,765 mention links extracted from 22,977 full-text articles published in 12 Anglophone philosophy journals between 1890 and 1979 and conduct a comparative analysis of these mentions over time. We (1) explore the relative impact of 11 prominent representatives of logical empiricism, (2) compare the movement's reception with a number of contemporaneous philosophical schools, (3) map its evolving place in the American philosophical landscape through a series of co-mention analyses, and (4) dissect logical empiricism's reception by focusing on the role of various journals and universities.

Before we proceed, a warning is in order. The aim of this paper is not to present a *competing* method for the study of the reception of logical empiricism. Nor is it our ambition to refute or verify existing perspectives on the development of twentieth-century philosophy. Though bibliometric analyses may help confirm or challenge received views and hidden assumptions, mention statistics alone do not suffice to overturn or validate qualitative historical research. Citations and mentions are useful but imperfect proxies for intellectual influence such that detailed historical knowledge remains needed to both generate and interpret the data. EDHIPHY was developed in close collaboration with historians of twentieth-century American, German, and French philosophy (see acknowledgements) and many of the choices we had to make in selecting, cleaning, disambiguating, and interpreting the data were informed by existing research. It is better, therefore, to view the present study as an attempt to infuse prevailing discussions with new data, allowing historians to explore different types of questions and generate new research ideas. Quantitative methods do not *replace* traditional approaches, they offer logical empiricist scholars an additional tool, helping them explore the movement's reception through a new, wide-angle lens.

7.2 Mention Extraction and Analysis

Bibliometric analyses of academic corpora typically rely on *citations* in mapping connections between scholars, publications, ideas, journals, and institutions. Citations are (relatively) standardized references to outside publications and thus an ideal tool to explore links between large numbers of books and articles. Unfortunately, citations are not useful instrument to study mid-twentieth-century philosophy. Academic papers from this period typically include few standardized citations and there are severe technical limitations to extracting them on a large scale. Up until the

1980s, very few articles included a list of references and most journals allowed different citation styles, sometimes even within a single paper or issue.[1]

EDHIPHY relies on an alternative method to map and analyze links between scholars and institutions in philosophy. While articles published before 1980 typically include few bibliographic citations, they usually *mention* dozens of philosophers, scientists and other persons of interest. Moritz Schlick's "Meaning and Verification" (1936), for example, does not include a single standardized reference but mentions dozens of people, including Richard Avenarius, P. W. Bridgman, Rudolf Carnap, Albert Einstein, David Hume, Immanuel Kant, C. I. Lewis, Ernst Mach, John Stuart Mill, Bertrand Russell, Arthur Schopenhauer, Herbert Spencer, Hans Vaihinger, and Ludwig Wittgenstein. Some of these philosophers and scientists are mentioned more than others (Lewis 23 times, Mill just once) indicating their varying importance to Schlick's argument. EDHIPHY includes mention data of 22,977 articles published in 12 Anglophone philosophy journals between 1890 and 1979 and can be used to (1) generate detailed mention statistics of specific philosophers and journals, (2) track broad changes in mention behavior over time, and (3) generate networks of co-mention relations to identify clusters of philosophers and scientists who are frequently mention together.

In building EDHIPHY, we developed a procedure to extract mentions from our corpus of full-text articles.[2] In short, we combined four existing databases—including a list of philosophers in WikiData—to create a dictionary of 44,376 philosophers. This dictionary was used as an entity ruler to identify which combinations of strings in the full-text articles potentially refer to a philosopher.[3] In addition, we created a list of 139,623 aliases (3.7 variants per philosopher on average) and added them to our dictionary to increase the probability of finding matches. Rudolf Carnap, for example, has five aliases in EDHIPHY, including 'R. Carnap', the common misspelling 'Rudolph Carnap', and simply 'Carnap' to make sure that the most common variants are picked up by the extraction algorithm.

The next step of the process was to develop a procedure to *disambiguate* mentions. For while it may be obvious to a historian that Schlick meant to refer to Clarence Irving Lewis in mentioning "Lewis" on page 347 of "Meaning and Verification", our extraction algorithm is unable to determine which of the 28 philosophers named 'Lewis' in the dictionary he was referring to. In order to disambiguate such mentions, we used a series of strategies. One of these strategies searches for unambiguous mentions elsewhere in the paper and assigns the

---

[1] An exploratory investigation of a random sample of 80 articles published in *Journal of Philosophy* and *Philosophical Review* between 1921 and 1960 (five articles per journal per decade) reveals that a fifth of the publications does not include a single bibliographic citation. Not one of the articles in the sample included a bibliography or list of references.

[2] The full-text articles were obtained from JSTOR through its Data for Research service (Burns et al. 2009). A Python script was used to clean the documents and prepare them for mention extraction. This script i.a. corrects small OCR errors, removes page numbers, rejoins words split by a line break, and eliminates author names (which would otherwise count as mentions).

[3] See Petrovich's et al. (2024) for a more detailed description of the extraction procedure.

corresponding philosopher to all ambiguous mentions containing the same string of letters. Because Schlick unambiguously mentions "C. I. Lewis" on page 343 of his paper (and none of the other 'Lewises' in the dictionary), all 23 mentions of 'Lewis' in the paper are linked to Clarence Irving Lewis. Another strategy makes use of co-mention frequencies. Even though Schlick never specifies whether the string 'Spencer' is a reference to 'Herbert Spencer' or to any of the other philosophers named Spencer in our dictionary, one can still determine that 'Herbert Spencer' is by far the most likely option if we take into account (1) the other, unambiguously mentioned philosophers in the article and (2) the co-mention frequencies between the different 'Spencers' and all other philosophers in the corpus. Combined, these and two other strategies helped us to successfully link 93% of the mentions to specific philosophers, with an estimated reliability of 82–91%.[4]

The present paper uses a subset of EDHIPHY to generate mention statistics and co-mention networks concerning the reception of logical empiricism in American philosophy. Concretely, we limit our study to the 10,190 articles published in six American journals—*Journal of Philosophy, Philosophical Review, The Monist, Philosophical Studies, Philosophy and Phenomenological Research*, and *Philosophy of Science*—between 1921 and 1979.[5] In addition to mention data, we will use two sets of additional data that are also included in EDHIPHY: (1) A list 8.940 American philosophy dissertations including information about author, university, and year of completion; (2) A list of 559 philosophy hires at the assistant, the associate and the full professor level at 11 prestigious philosophy departments (Berkeley, Chicago, Columbia, Cornell, Harvard, Michigan, Pennsylvania, Princeton, Stanford, UCLA, and Yale) between 1930 and 1970.[6] This additional data is used to group mention statistics by author affiliation (analysis 4).

7.3 Analysis 1: Eleven Logical Empiricists

Figure 7.1 presents the mention statistics of 11 philosophers, logicians and scientists commonly associated with the logical empiricist movement: A. J. Ayer, Gustav Bergmann, Rudolf Carnap, Herbert Feigl, Philipp Frank, Carl Gustav Hempel, Otto Neurath, Karl Popper, Hans Reichenbach, Moritz Schlick, and Alfred Tarski. It displays the number of total mentions to each of the 11 empiricists (top graph), the number of distinct articles mentioning them at least once (middle graph), and the

[4] The reliability of the links was estimated through a manual assessment of a random sample of one thousand mentions. See Petrovich et al. (2024) for a description of the two other disambiguation strategies and the reliability test.

[5] Ocassionally, we will compare the US and the British contexts. In these analyses we also make use of the British journals in the database: *Analysis, Mind, Philosophical Quarterly*, *Philosophy*, and *Proceedings of the Aristotelian Society*. Note that not all journals had been founded in 1921, such that analyses of different decades will be based on different sets of journals.

[6] http://www.proquest.com; Strassfeld (2020).

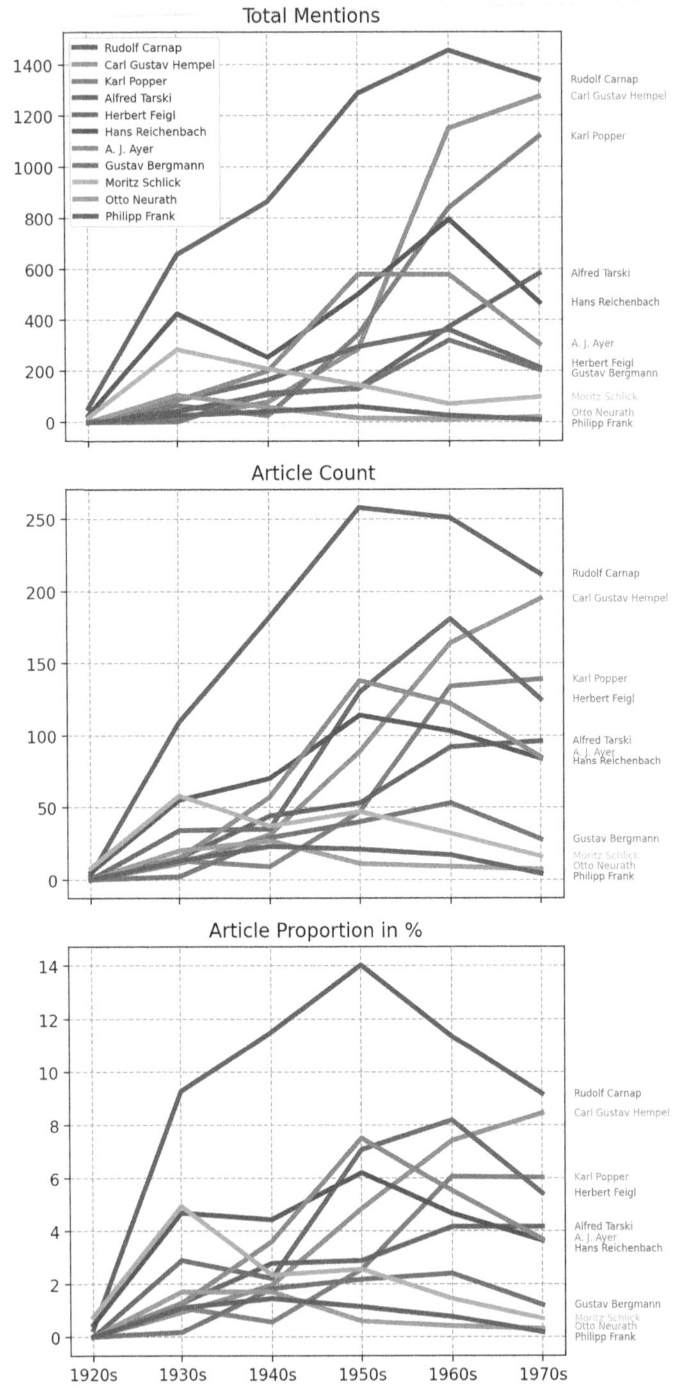

Fig. 7.1 Mention statistics of 11 logical empiricists in American philosophy journals by decade, displaying (**a**) total number of mentions, (**b**) number of distinct mentioning articles, (**c**) proportion of distinct mentioning articles

proportion of mentioning articles mentioning them at least once (bottom graph) between the 1920s and the 1970s, thereby offering three measures to estimate their relative impact and to track shifts in their influence over time.[7]

The three graphs reveal that Carnap was the most-mentioned logical empiricist in every decade between 1930 and 1980. Likely, this will not be a surprise given the impact of his *Der Logische Aufbau der Welt* (1928) and *Logische Syntax der Sprache* (1934) as well as the influence of the so-called Carnap-Quine debate in the development of analytic philosophy. More remarkable is the magnitude of Carnap's influence. At the height of his fame in the 1950s, a stunning 14.1% of all articles in our six American journals mention Carnap at least once (with an average of 5 mentions per mentioning article). His impact is also impressive if we compare it to the other logical empiricists. In the 1940s, approximately 42% of all mentions connected to one of the aforementioned logical empiricists were mentions of Carnap.

Reichenbach is, overall, the second most-mentioned logical empiricist between 1921 and 1950, reflecting the influence of his work in the philosophy of physics, probability, and the popularity of textbooks such as *Experience and Prediction* (Reichenbach 1938). His influence starts to decline after his untimely death in 1953 and his second-place position is taken over by Ayer, whose numbers more than double in the 1950s, echoing the impact of, among others, his notorious *Language, Truth, and Logic* (Friedman 1999). Even more remarkable is the growth of Feigl's influence. His mention numbers quintuple between the 1940s and the 1960s, making him the second-most mentioned logical empiricist of the latter decade (in terms of distinct mentioning articles). Feigl's impact likely reflects his prominent role in debates about meaning, scientific realism, and the mind-body problem as well as his active role in the production of anthologies (Feigl and Sellars 1949; Feigl and Brodbeck 1953) and editor of volumes of the *Minnesota Studies in Philosophy of Science* series (e.g. Feigl and Scriven 1956).[8] By the 1970s, Hempel becomes the second most-mentioned logical empiricist, just before Popper, Feigl, and Tarski, demonstrating the growing importance of these second-generation logical empiricists. As yet, little work has been done on the American reception of Popper and Tarski, whose numbers continue to grow in a period when references to other second-generation logical empiricists (e.g. Feigl, Ayer, and Bergmann) start to decline, suggesting that there is an interesting story to tell about the reception of their work.

[7] The number of total mentions is equal to or greater than the total number of distinct mentioning articles because articles can mention a philosopher more than once. Schlick, for example, is mentioned 825 times in 198 distinct articles, meaning that he is mentioned 4.17 times per mentioning article on average.

[8] The growth of Feigl's mention numbers is less impressive if we focus on total mentions. This is likely an effect of the popularity of his aforementioned anthologies. These are mentioned in many distinct articles but are rarely discussed in-depth. In the 1960s, Feigl's is mentioned just twice per mentioning article while Reichenbach averages almost 8 mentions per mentioning article in the same period (see footnote 7).

While Fig. 7.1 displays some (perhaps) unanticipated results concerning the magnitude of Carnap's influence (in the 1940s and 1950s) and the reception of Popper, Feigl, and Tarski (in the 1960s and 1970s), it confirms existing work on the marginalization of the perspectives of Neurath and Frank, who appear to have played little to no role in American philosophical debates (see Sect. 7.1). Figure 7.2 presents the proportion of references to each of the 11 logical empiricists as a percentage of the combined number of mentions and shows that both Frank and Neurath received no more than 1% of these mentions each. Figure 7.3, finally, displays the proportion of mentions of each of the 11 logical empiricists in five *British* journals of philosophy (see footnote 5), showing that the two were (almost) equally 'unpopular' in the United Kingdom. A comparison of Figs. 7.2 and 7.3, finally, reveals that country of residence was an important factor in logical empiricism's reception in the Anglophone world. Ayer and Popper (both based in the UK) receive a much larger share of the mentions in British than in American journals. The converse holds for Carnap, Feigl, Hempel, Reichenbach, and Bergmann (all based in the US).[9]

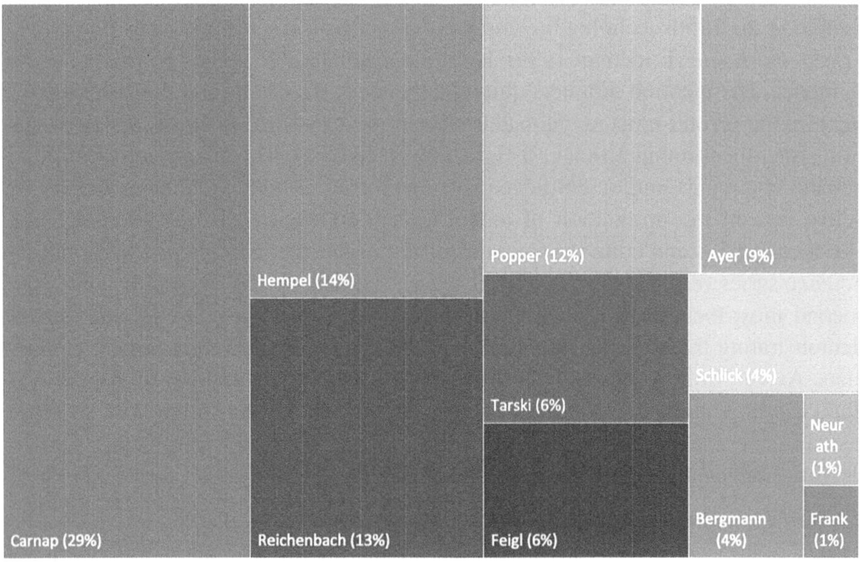

Fig. 7.2 Proportion of mentions (article count) of 11 logical empiricists in American philosophy journals between 1921 and 1979

[9] Considering this last conclusion, one may expect Friedrich Waismann to have done rather well in the UK, too. This is incorrect. Waismann received fewer mentions than Neurath, both in British and in American journals.

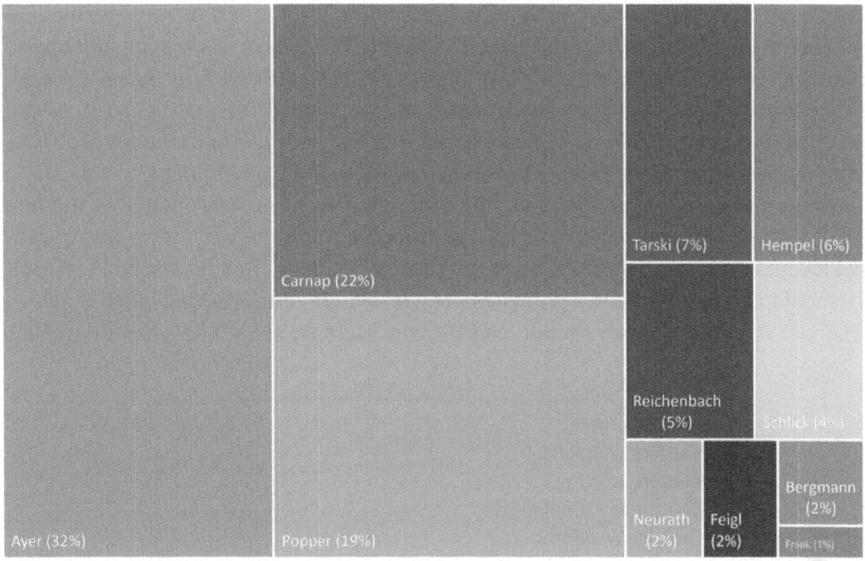

Fig. 7.3 Proportion of mentions (article count) of 11 logical empiricists in British philosophy journals between 1921 and 1979

Table 7.1 Most-mentioned contemporary authors (article count) in American philosophy journals by decade

1921-30		1931-40		1941-50		1951-60		1961-70		1971-79	
James	242	Whitehead	254	Dewey	314	Dewey	290	Wittgenstein	307	Quine	376
Russell	209	Dewey	239	Whitehead	253	Russell	286	Quine	273	Davidson	294
Bergson	177	Russell	213	Russell	240	Carnap	258	Russell	252	Wittgenstein	227
Whitehead	146	James	179	James	226	Whitehead	209	Carnap	251	Carnap	212
Dewey	146	Einstein	157	Carnap	182	James	179	Feigl	181	Harman	197
Santayana	131	Bergson	141	Bergson	145	Quine	168	Moore	171	Hempel	195
Einstein	130	Santayana	123	C. I. Lewis	138	Moore	167	Ryle	167	Frege	191
Bradley	118	Bradley	120	Husserl	126	C. I. Lewis	163	Hempel	164	Chisholm	183
Bosanquet	113	C. I. Lewis	118	Peirce	126	Wittgenstein	161	Strawson	151	Russell	177
Spencer	98	Carnap	109	Santayana	116	Peirce	156	Dewey	144	Strawson	168
Royce	90	Peirce	100	Moore	116	Einstein	141	Whitehead	143	Goodman	156
Broad	64	Wittgenstein	87	Einstein	110	Ayer	138	Goodman	142	Putnam	154
Alexander	62	Eddington	81	Mead	81	Feigl	130	Einstein	137	Hintikka	151
Eddington	61	Lovejoy	78	Heidegger	75	Bergson	119	Popper	134	W. Sellars	150
Morgan	55	Bosanquet	73	Nagel	72	Reichenbach	114	Black	130	Kripke	146
Holt	51	Husserl	70	Reichenbach	70	Ryle	111	Frege	130	D. Lewis	142
Wundt	48	Moore	68	Bradley	69	W. Sellars	111	W. Sellars	126	Popper	139
Lovejoy	46	Mach	65	Morris	65	Broad	109	Ayer	122	Feigl	125
Croce	46	Royce	61	Wittgenstein	62	Santayana	107	James	119	Lehrer	125
Mach	45	Bridgman	61	Broad	61	Black	105	Maxwell	119	Black	123

7.4 Analysis 2: Philosophical Movements

Carnap was not just one of the most-mentioned logical empiricists. Table 7.1 lists the top-20 most-mentioned contemporary philosophers and scientists per decade (article count), revealing that he was among the most influential academics in

American philosophy more generally.[10] He enters the top ten of most-mentioned authors in the 1930s and is firmly located within the top five in the four subsequent decades. Nor is he the only logical empiricist to be included in this list. Feigl, Reichenbach, Ayer, Popper, and Hempel are included in several decades, as are some of the young American philosophers—e.g. Ernest Nagel and Charles Morris—who played a role in the promotion of their work in the United States.

Since most logical empiricists only emigrated to the United States in the late 1930s, their omnipresence in U.S. philosophical discussions is a remarkable achievement. If we compare the reception of Carnap, Reichenbach, Feigl and Hempel to other well-known philosophers who sought refuge in the United States—Theodor Adorno, Hannah Arendt, and Herbert Marcuse (Fig. 7.4)—the differences

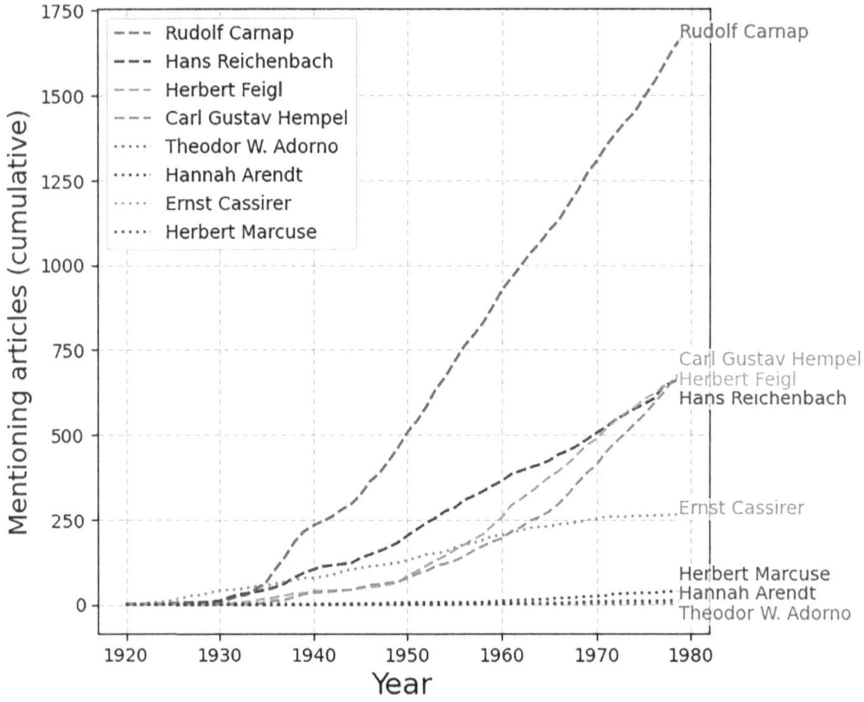

Fig. 7.4 Cumulative mentions (article count) of eight European migrants

[10] Table 7.1 only includes academics who were alive on or born after January 1, 1901, excluding historical figures such Hume and Hegel. If we had included them, James would have been the fourth most-mentioned philosopher of the 1920s, after Kant (335 mentions), Plato (286), and Aristotle (270). In addition, Ullin Place and Richard J. Blackwell were removed from the top lists of the 1920s and 1970s respectively because their rankings are the result of incorrect mention identifications. Most occurrences of the string 'Blackwell' are actually references to the publisher; most occurrences of the string 'Place' are actually uses of the English term 'place'. See Petrovich et al. (2024) for a detailed discussion of the incidence of false positives in EDHIPHY.

are striking. Only Ernst Cassirer had a reception comparable to, e.g. Feigl and Reichenbach, up until his death in 1945. In the 1940s, approximately 20% of all articles published in six American journals mention at least one logical empiricist. A decade later, when the movement was at the peak of its influence, this number grows to about one in three.

Table 7.1 can also help suggest developments that paved the way for the reception of logical empiricism. While American philosophers appear to have largely ignored logical empiricists in the 1920s (see Fig. 7.1), the list of most-mentioned authors in this period reveals that debates about the foundations of relativity theory (e.g. Albert Einstein and Arthur Eddington) and psychology (e.g. Wilhelm Wundt, E. B. Holt, and C. Lloyd Morgan) significantly affected the development of US philosophy (cf. Verhaegh 2024a, b). Bertrand Russell, well known for his work in logic and the foundations of mathematics as well as the promotion of "scientific philosophy" in his 1914 Harvard Lectures, is one of the most-mentioned philosophers in 1920s and 1930s. As was A. N. Whitehead, who was best-known for the *Principia Mathematica* and his work in philosophy of science up until 1924, when he accepted a position at Harvard.[11] Within philosophy proper, several representatives of the Cambridge school of analysis—e.g. C. D. Broad, G. E. Moore, and Ludwig Wittgenstein—are included in the list of most-mentioned philosophers. All these philosophers and scientists had direct or indirect ties with the logical empiricists, which may help explain why the US was fertile soil for their perspective and approach.

Finally, Table 7.1 can help us estimate when and to what degree certain (competing) schools and movements played a role in American philosophy. Several prominent representatives of the (British) idealist movement—e.g. F. H. Bradley, Bernard Bosanquet, Josiah Royce, and Benedetto Croce—are among the most-mentioned philosophers of the 1920s but their mention numbers rapidly decline in subsequent decades. Something similar applies to classic American pragmatists such as William James, C. S. Peirce, and John Dewey. Much has been said about the decline of pragmatism in the wake of the analytic turn, but these numbers can help estimate when and to what degree this happened.[12] Figure 7.5 offers a preliminary way to visualize some of these effects. It compares the total mentions of the top-3 most-mentioned pragmatists, British analysts, idealists, logical empiricists, U.S analytic philosophers, and Bergsonists/phenomenologists and tracks the growth and/or decline of these proportions over time.[13] In addition to the decline of the classical pragmatist

[11] After the publication of *Process and Reality*, Whitehead also became known for his metaphysics, which makes it more difficult to interpret his mention statistics from the 1930s onwards.

[12] Rorty (1995) and Misak (2013) represent various perspectives on the development of pragmatism. We will return to this issue in analysis 3 below.

[13] Depending on the decade, these top 3s include James, Santayana, Dewey, Peirce, and C. I. Lewis (pragmatism), Russell, Wittgenstein, Moore, Broad, Ryle, and Strawson (British analytic philosophy), Bradley, Bosanquet, Royce, and T. H. Green (idealism), Carnap, Reichenbach, Schlick, Tarski, Hempel, and Feigl (logical empiricism), Quine, Stevenson, W. Sellars, Goodman, Davidson, and Harman (American analytic philosophy), and Bergson, Husserl, Hartmann, Heidegger, and

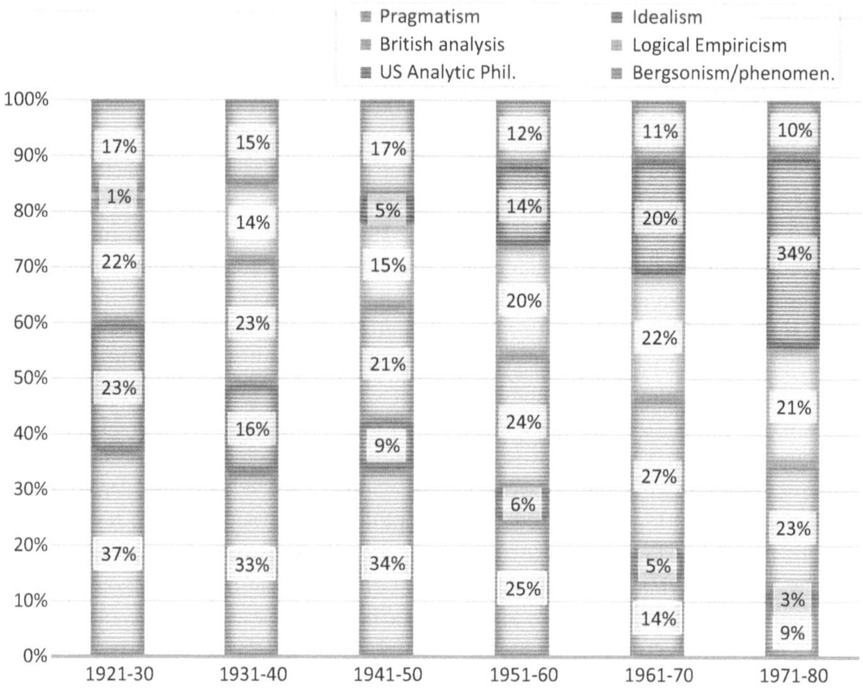

Fig. 7.5 Proportion of mentions of pragmatists, idealists, British analysts, logical empiricists, American analytic philosophers, and Bergsonists/phenomenologists by decade based on the top-3 most-mentioned philosophers of each movement per decade

and idealist movements, this figure showcases the explosive growth of American analytic philosophy from the 1940s onwards. Logical empiricism, on the other hand, encompasses a rather stable proportion of mentions between the 1950s and the 1970s, despite or perhaps because of the rapid growth of the analytic movement.

7.5 Analysis 3: Co-mention Networks

While Fig. 7.5 can help us track the development of various school and movements, it makes use of external, historically contentious labels such as 'pragmatism' and 'idealism' to group philosophers together. Such school labels are problematic since (1) there is no clear-cut, generally agreed upon definition of when someone does or does not qualify as a 'phenomenologist' or 'logical empiricist', (2) they are

Sartre (Bergsonism/phenomenology). We excluded Whitehead from these top-3 s since it is difficult to disentangle references to Whitehead qua logician and Whitehead qua process philosopher. See footnote 11. We also excluded Ayer because he played a prominent role in debates about logical empiricism and British analytic philosophy more generally.

relatively inflexible, making it difficult to do justice to philosophers whose work is associated with various approaches in different periods of their careers (e.g. A. J. Ayer or Richard Rorty), and (3) they fail to incorporate shifts in meaning regarding these labels themselves. 'Analytic philosophy', for example, meant something different when Nagel wrote "Impressions and Appraisals of Analytic Philosophy in Europe" (1936) than it did when Robert Ammerman published *Classics of Analytic Philosophy* (1965).

An alternative method to map connections between philosophers is to track their *co-mention* relations. Since there are many more articles that mention both Hempel and Nagel than articles mentioning both Hempel and Sartre (even though Nagel and Sartre have comparable mention numbers), one can conclude that the former were perceived to be doing more similar work than the latter. A co-mention *network* visually represents co-mention relations between all mentioned authors in a corpus and clusters them into groups of authors who are frequently mentioned together. Co-mention networks, unlike school labels, do not rely on external classifications, they are multi-dimensional because they represent a philosopher's relation with every other author in a community, and they are sensitive to historical shifts because one can track developments within a corpus by comparing co-mention networks representing distinct periods of publication. Figures 7.6, 7.7, 7.8 and 7.9 present co-mention networks of the (approximately) top-200 most-mentioned authors in American philosophy in four decades: the 1920s, the 1930s, the 1950s, and the 1970s. Each node in these networks represents a particular philosopher or scientist, the sizes of the nodes represent their mention numbers (article count), and the location of each node reflects the co-mention distance between the author and all other authors in the network.[14]

Figure 7.6 presents a co-mention network of the American philosophical literature in the 1920s. It is divided into five clusters. The largest cluster is a group of mostly Anglophone philosophers located on the left (the blue cluster). This group is sandwiched between a red cluster comprising mostly German-speaking philosophers and a green cluster comprising mostly logicians and scientists. Philosophers associated with the pragmatist and realist traditions (e.g. Dewey, James and Santayana) are located near the logicians and the scientists in the green cluster, while philosophers commonly associated with the idealist tradition (e.g. Bradley, Bosanquet, and Royce) are nearer to the German philosophers in the red cluster.[15] C. I. Lewis and Whitehead are some of the only philosophers located within the green cluster (positioned on the border between the blue and the green cluster),

[14] The networks are generated by VOSviewer. See Waltman et al. (2010) and Van Eck and Waltman (2010) for expositions of VOSViewer's mapping and clustering algorithms. In all maps, we used the default clustering resolution (1.0). The minimum cluster size was set to 20. Authors whose inclusion is mostly the result of incorrect mention identifications (footnote 10) were removed.

[15] Not all these names are visible in Fig. 7.6 which presents a screenshot of the network. A complete interactive version of the network can be accessed via the link in the Figure caption. The same applies to Figs. 7.7, 7.8 and 7.9 below. Some of the clusters in the online networks have different colors but they are the same in every other respect.

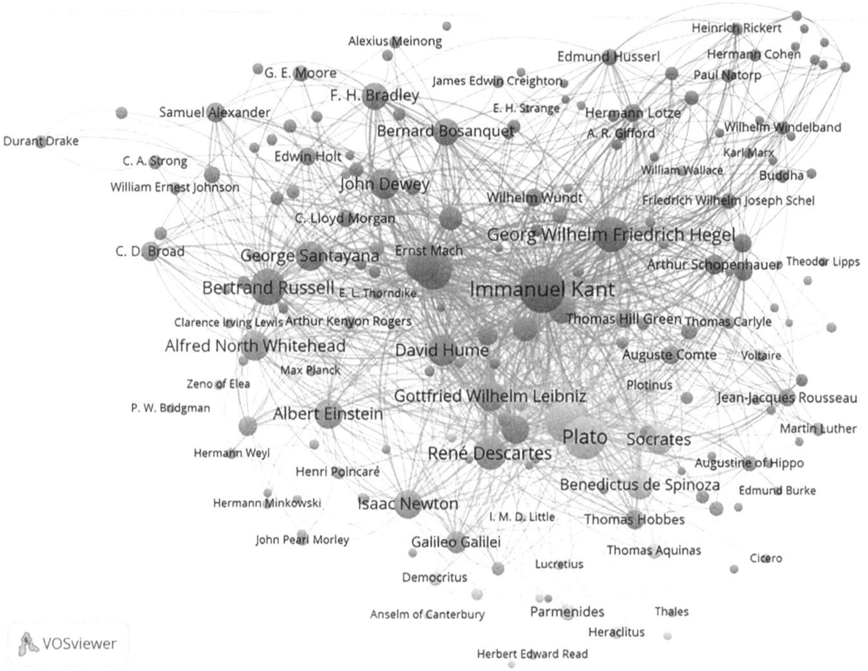

Fig. 7.6 Co-mention network of most-mentioned philosophers 1921–1930. (Complete interactive network available at https://tinyurl.com/2chlfo78)

reflecting their contributions to the development of symbolic logic and debates about the philosophical implications of relativity theory. The yellow and purple clusters, finally, contain mostly historical figures and historians of philosophy. The yellow cluster contains mostly ancient and medieval philosophers, the purple cluster mostly (early) modern philosophers.

Figure 7.6 presents the American philosophical landscape before the intellectual migration. Figure 7.7 presents a co-mention network of the 1930s and can be used to study how the rise of logical empiricism changed U.S. philosophy. The philosophers and scientists in the network are divided into roughly the same groups, except for the two historical clusters which have merged into a single cluster. Yet we can clearly detect an expansion of the green cluster, which now comprises a substantial group of scientists, logicians, *and* philosophers, including all logical empiricists (i.e. Carnap, Reichenbach, Schlick, Feigl, and Neurath).[16] Interestingly, this network also comprises (a) (former) Cambridge analysts who contributed to the development of mathematical logic (e.g. Russell, Whitehead, and Ramsey) but *not* their Cambridge colleagues who relied on a more ordinary language approach (e.g. Moore and Broad); (2) pragmatists who defended an anti-psychologist perspective

[16]Again, not all these names are visible in Fig. 7.7. See footnote 15.

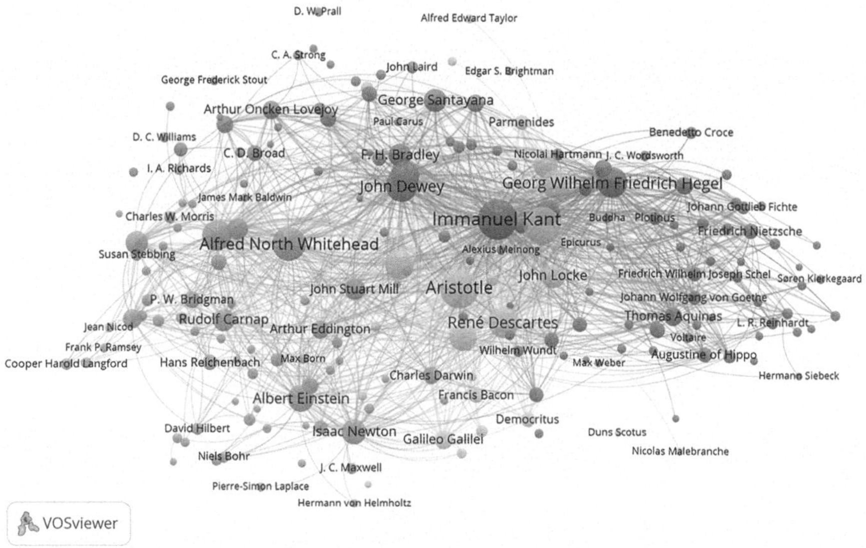

Fig. 7.7 Co-mention network of most-mentioned philosophers 1931–1940. (Complete interactive network available at https://tinyurl.com/2y59rt7q)

on logic and used formal tools in their research (e.g. Peirce and Lewis) but not their more naturalistically oriented colleagues (e.g. Dewey, James, and Santayana); and (3) representatives of a new generation of American philosophers who were interested in scientific philosophy and semiotics (e.g. Nagel and Morris) but not their contemporaries who advocated a different approach (e.g. D. C. Williams and A. E. Murphy). The division between the blue and green cluster, in sum, cuts across now common labels such as 'pragmatism' and 'analytic philosophy', suggesting that the divide between formal/anti-psychologistic and non-formal/naturalistic philosophy played a more significant role at the time.

Figure 7.8 portrays the American philosophical landscape in the 1950s. The logical empiricists are still located in a (green) cluster comprising mostly scientists, logicians, and philosophers of science but most other philosophers (including Lewis, Wittgenstein, and Whitehead) have 'moved backed' into the region where most Anglophone philosophers are located. This region is now split into two distinct groups: a blue cluster comprising mostly analytic philosophers (e.g. Quine, Moore, and Sellars) and a turquoise cluster including an amalgam of non-analytic American philosophers—e.g. process philosophers such as Paul Weiss and Charles Hartshorne, naturalists such John Herman Randall Jr. and Roy Wood Sellars as well as some major figures in the history of American philosophy (e.g. James, Peirce, and Royce). Bergman and Reichenbach are firmly positioned within the green cluster, while Carnap and Feigl are located near the edge, close to e.g. Russell and Quine, reflecting their role in discussions in analytic philosophy proper (e.g. the Carnap-Quine debate about analyticity and Feigl's contribution the philosophy of mind).

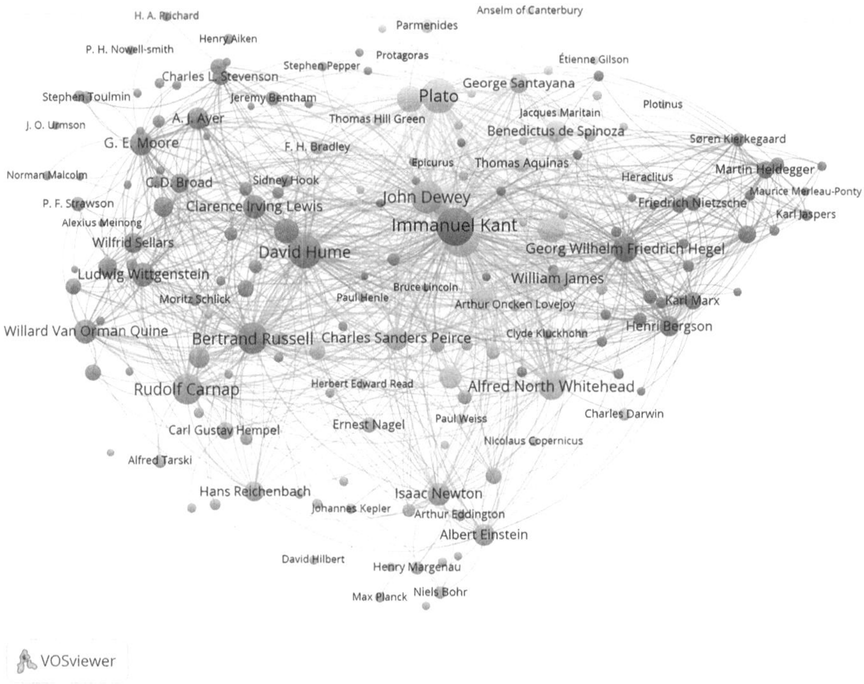

Fig. 7.8 Co-mention network of most-mentioned philosophers 1951–1960. (Complete interactive network available at https://tinyurl.com/2ax79cbn)

The gradual separation between philosophy of science and philosophy proper is even more evident in Fig. 7.9, which presents the American landscape in the 1970s. In this period, analytic philosophy has come to play such a dominate role that all four clusters contain philosophers working in the analytic tradition. While historical figures (e.g. Aristotle and Hume) and representatives of various schools of American (e.g. James and Dewey) and continental philosophy (e.g. Husserl and Heidegger) were located in separate clusters up until the 1950s, not even the combination of the three is sizeable enough to form a separate cluster by the 1970s. The three groups are cramped together into one (red) cluster which also contains a variety of analytic ethicists and political philosophers (e.g. Nozick, Hampshire, Feinberg, and Rawls). The blue cluster comprises mostly technical analytic philosophers who contributed to logic, metaphysics, and philosophy of language (e.g. Davidson, Geach, Hintikka, David Lewis, and Quine), while the turquoise cluster comprises mostly analytic philosophers with a less technical approach, including most ordinary language philosophers (Austin and Ryle) as well as various epistemologists and philosophers of mind and action (Anscombe, Gettier, and Malcolm). Within the green cluster, finally, one can detect various approaches and subgroups, too: Logical empiricists who emphasized philosophy's role in studying the logic of science (e.g. Carnap, Hempel, and Suppes) are located at the top of the cluster near prominent

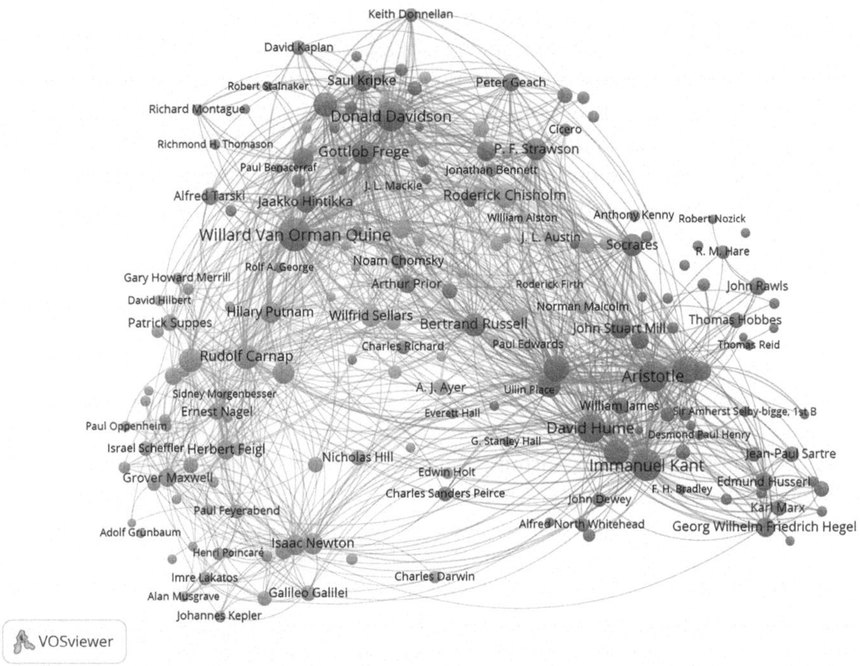

Fig. 7.9 Co-mention network of most-mentioned philosophers 1971–1979. (Complete interactive network available at https://tinyurl.com/2xjsec4r)

philosopher-logicians such as Putnam and Ramsey, while philosophers of science experimenting with historical approaches and post-positivist perspectives (e.g. Feyerabend, Hanson, Lakatos, and Kuhn) are located near the bottom.

7.6 Analysis 4: Journals and Institutions

All in all, the co-citation networks in Figs. 7.6, 7.7, 7.8 and 7.9 suggest that logical empiricism had a remarkably consistent and stable position within American philosophy—an intellectual niche of logicians, scientists, and scientific philosophers that already existed in the 1920s, before the logical empiricists played a role in US debates. It is important to note, however, that these results partly reflect our selection of journals. If one excludes specialist philosophy of science journals such as *Philosophy of Science* and *The Monist*, the green and blue clusters collapse into a single cluster in some (but not all) decades, such that logical empiricism seems more firmly located within analytic philosophy. This effect reflects the dual nature of logical empiricism's reception in North America. For most of its history, logical empiricism was simultaneously treated as a movement in analytic philosophy and in scientific philosophy (later: philosophy of science) broadly conceived.

In order to explore the role of various journals in some more detail, Table 7.2 lists the top-15 most-mentioned philosophers in three journals in the 1950s, showing that these periodicals had remarkably different profiles in this period. While *Journal of Philosophy*'s top 15 mostly consists of historical figures and major pragmatists (Dewey, James, Lewis, Peirce, and Santayana), articles published in *Philosophical Studies* focused on an entirely different range of authors. Its top 15 includes very few historical figures (only Hume and Kant) and all contemporary philosophers belong to the analytic tradition broadly construed. Analyses of the same journals in the 1960s and 1970s reveal that *Journal of Philosophy* would soon move into a similar direction (see also Katzav 2018), suggesting that *Philosophical Studies*—a journal founded by a logical empiricist—was an important trend setter in this regard. *Philosophy of Science*, finally, had a still different profile. Its top-15 includes a mix of historical philosophers and scientists (e.g. Newton and Galileo), philosophers of science in various traditions (e.g. Reichenbach, Margenau, and Whitehead) as well as American pragmatists, confirming the aforementioned bifurcation between analytic philosophy and philosophy of science.

EDHIPHY can also be used to explore the role different *departments* of philosophy played in the reception of logical empiricism. Table 7.3 presents the most-mentioned authors in articles authored by philosophers affiliated with three departments of philosophy—Columbia, Princeton, and UC Berkeley—between 1951 and 1960.[17] It reveals that these departments had rather different profiles, too. While philosophers at Columbia—the department of Dewey and his naturalist

Table 7.2 Top-15 most-mentioned philosophers in Journal of philosophy, Philosophical studies, Philosophy of science, 1951–1960. Historical figures are marked ***

Journal of philosophy		Philosophical studies		Philosophy of science	
***Kant	157	Carnap	46	Einstein	73
Dewey	149	Quine	32	***Newton	65
***Plato	146	Feigl	29	Carnap	60
***Aristotle	141	W. Sellars	29	***Kant	56
***Hume	139	Russell	29	***Aristotle	49
Russell	93	Black	19	***Plato	42
Whitehead	86	Reichenbach	18	Margenau	40
***Socrates	85	Wittgenstein	18	Russell	40
James	78	***Hume	18	***Hume	36
***Mill	76	Broad	18	***Galileo	35
***Hegel	75	Moore	17	Whitehead	34
***Descartes	72	Frege	14	Dewey	34
Peirce	72	Goodman	13	Reichenbach	33
Santayana	67	***Kant	13	Feigl	32
C. I. Lewis	63	Ayer	13	James	30

[17] An article published in year *y* was assigned to a department *d* iff (a) one of the authors was employed by *d* in *y* or *y*-1 or (b) one of the authors obtained their Ph.D. at *d* between *y*-3 and *y* + 3.

Table 7.3 Top-10 most-mentioned philosophers by philosophers affiliated with Berkeley, Columbia, and Princeton, 1951–1960. Historical figures are marked ***

Columbia Univ.		Princeton Univ.		UC. Berkeley	
Dewey	20	Carnap	18	Carnap	9
***Plato	19	***Hume	10	Russell	6
***Aristotle	15	Goodman	9	Frege	6
***Kant	15	Quine	9	Wittgenstein	4
***Hume	13	Russell	8	Strawson	4
***Mill	11	***Plato	8	Quine	4
***Socrates	10	Feigl	8	Einstein	4
***Hegel	9	Wittgenstein	7	Tarski	3
***Descartes	8	***Socrates	7	Church	3
Russell	8	***Aristotle	7	Moore	3

students—predominantly mentioned their teacher and a series of historical figures, their colleagues at Berkeley were mentioning mostly analytic philosophers (broadly conceived). The logical empiricists were particularly well received at Princeton. Carnap, Feigl, Hempel, and Reichenbach are all included in the top-25 most-mentioned philosophers, reflecting the department's shifting focus in the 1950s, when it hired Kemeny (1952), Putnam (1953), and Hempel (1955) and started to produce some more technically oriented graduate students (e.g. Nicholas Rescher, Richard Jeffrey, and Paul Benacerraf).

Naturally, these departmental profiles are the result of a variety of faculty and graduate students citing different types of philosophers. These internal differences can be studied with EDHIPHY as well. Table 7.4 presents an analysis of Yale's department of philosophy in 1963, a period when its faculty started to become heavily split (see Kuklick 2004). We have analyzed which professors and graduate students cited some of the most-mentioned logical empiricists (Carnap, Feigl and Bergmann), pragmatists (Dewey, James, and Peirce), and phenomenologists/existentialists (Heidegger, Husserl, and Sartre), indicating which people belonged to which camp. In addition to large-scope analyses of American philosophy on the most general level, in sum, EDHIPHY can also assist historians in conducting more fine-grained reconstructions.

7.7 Conclusion

Logical empiricism played an important role in the development of American philosophy. This paper has tried to map its reception by quantitatively analyzing 10,190 articles published in six American journals between 1921 and 1979. We (1) explored the relative impact of 11 logical empiricists, (2) compared their influence with a number of contemporaneous philosophical movements, (3) mapped its evolving

Table 7.4 Mention-based analysis (1961–1970) of philosophers affiliated with Yale in 1963. Only faculty and graduate students (*) mentioning at least three of the nine philosophers are included.

	Carnap	Feigl	Bergmann	Dewey	James	Peirce	Heidegger	Husserl	Sartre
John Bacon*	X	X	X						
Henry Margenau	X		X		X				
Wilfrid Sellars	X	X	X			X			
Richard Bernstein				X	X	X	X		
John E. Smith	X			X	X	X			
Rulon Wells			X	X		X			
Kenneth Megill*						X	X		X
George Schrader				X			X	X	X
John D. Wild		X			X		X	X	X

place in American philosophy through a series of co-mention analyses, and (4) explored its reception on a more fine-grained level by analyzing differences between various journals and departments of philosophy.

Naturally, our maps and statistics depend on a number of contingent choices. The mention rankings and co-mention networks presented in Figs. 7.1, 7.2, 7.3, 7.4, 7.5, 7.6, 7.7, 7.8, and 7.9 and Tables 7.1, 7.2, 7.3, and 7.4 would have looked different if we had decided to include book publications, reviews, and APA conference proceedings; or if we had selected a different set of journals. In this sense, quantitative historical research is not fundamentally different from traditional, qualitative inquiry, in which historians continuously face choices concerning the selection of sources, too. On the more technical side, we had to make various decisions regarding data cleaning, mention extraction, and mention disambiguation. In making these decisions, we always tried to put historical considerations first. EDHIPHY was developed in collaboration with historians of twentieth-century American, German, and French philosophy in order ensure that such choices were informed by existing research. In combining technological innovation with historiographical depth, we aimed to created an interesting new tool to explore different types of questions and generate new research ideas. The present paper took some first steps toward exploring the potential of EDHIPHY (and mention analysis more generally) concerning the study of the American reception of logical empiricism. We hope it will stimulate historians to develop more detailed analyses in the future.

Acknowledgments EDHIPHY was developed as part of the project "Exiled Empiricists: American Philosophy and the Intellectual Migration", funded by the European Research Council (ERC, grant 2021-STG, 101039904) and the Dutch Research Council (NWO, grant VI. Vidi.201.115). All authors of this article are members of the project team. We like to thank the other team members—Claudia Cristalli, Fons Dewulf, Nina IJdens, and Ties van Gemert—for their help with the development of EDHIPHY and their comments on several versions of this article.

References

Ammerman, Robert R. 1965. *Classics of Analytic Philosophy*. New York: McGraw-Hill.

Braat, Michiel, Jan Engelen, Ties van Gemert, and Sander Verhaegh. 2020. The Rise and Fall of Behaviorism: The Narrative and the Numbers. *History of Psychology* 23 (3): 252–280.

Burns, John, Alan Brenner, Keith Kiser, Michael Krot, Clare Llewellyn, and Ronald Snyder. 2009. JSTOR: Data for Research. In *Research and Advanced Technology for Digital Libraries*, ed. Maristella Agosti, S. José Borbinha, Christos Papatheodorou Kapidakis, and Giannis Tsakonas. Berlin: Springer.

Carnap, Rudolf. 1928. *Der logische Aufbau der Welt*. Berlin: Weltkreis-Verlag.

———. 1934. *Logische Syntax der Sprache*. Vienna: Springer.

Dewulf, Fons. 2021. The Institutional Stabilization of Philosophy of Science and its Withdrawal from Social Concerns after the Second World War. *British Journal for the History of Philosophy* 29 (5): 935–953.

Feigl, Herbert, and May Brodbeck. 1953. *Readings in the Philosophy of Science*. New York: Appleton-Century-Crofts, Inc.

Feigl, Herbert, and Michael Scriven. 1956. *The Foundations of Science and the Concepts of Psychology and Psychoanalysis*. Minneapolis: University of Minnesota Press.

Feigl, Herbert, and Wilfrid Sellars. 1949. *Readings Philosophical Analysis*. New York: Appleton-Century-Crofts, Inc.

Friedman, Michael. 1999. *Reconsidering Logical Positivism*. Cambridge: Cambridge University Press.

Giere, Ronald. 1996. From *Wissenschaftliche Philosophie* to Philosophy of Science. In *Origins of Logical Empiricism*, ed. Ronald Giere and Alan W. Richardson, 335–354. Minneapolis: University of Minnesota Press.

Hardcastle, Gary L., W. Alan, and Richardson. 2003. *Logical Empiricism in North America*. Minneapolis: University of Minnesota Press.

Howard, Don A. 2003. Two Left Turns Make a Right: On the Curious Political Career of North American Philosophy of Science at Midcentury. In *Logical Empiricism in North America*, ed. Gary L. Hardcastle and Alan W. Richardson, 25–93. University of Minnesota Press.

Jewett, Andrew. 2011. Canonizing Dewey: Naturalism, Logical Empiricism, and the Idea of American Philosophy. *Modern Intellectual History* 8 (1): 91–125.

Katzav, Joel. 2018. Analytic Philosophy, 1925–1969: Emergence, Management and Nature. *British Journal for the History of Philosophy* 26 (6): 1197–1221.

Kuklick, Bruce. 2004. Philosophy at Yale in the Century after Darwin. *History of Philosophy Quarterly* 21 (3): 313–336.

Limbeck-Lilienau, Christoph. 2012. Carnap's Encounter with Pragmatism. In *Rudolf Carnap and the Legacy of Logical Empiricism*, ed. Richard Creath, 89–111. Dordrecht: Springer.

Misak, Cheryl. 2013. *The American Pragmatists*. Oxford: Oxford University Press.

Nagel, Ernest. 1936. Impressions and Appraisals of Analytic Philosophy in Europe. I. *The Journal of Philosophy* 33 (2): 29–53.

Petrovich, Eugenio, Sander Verhaegh, Gregor Bös, Claudia Cristalli, Fons Dewulf, Ties van Gemert, and N. IJdens. 2024. Bibliometrics Beyond Citations: Introducing Mention Extraction and Analysis. *Scientometrics* 129: 5731–5768. https://doi.org/10.1007/s11192-024-05116-x.

Reisch, George A. 2005. *How the Cold War Transformed Philosophy of Science. To the Icy Slopes of Logic*. Cambridge: Cambridge University Press.

Reisch, George A. and Adam Tamas Tuboly., manuscript. *The Road Not Taken: Philipp Frank on the Integration of Science, Facts, and Values.*

Reichenbach, Hans. 1938. *Experience and Prediction*. Chicago: University of Chicago Press.

Richardson, Alan W. 1998. *Carnap's Construction of the World*. Cambridge: Cambridge University Press.

———. 2017. From Scientific to Analytic: Remarks on How Logical Positivism Became a Chapter of Analytic Philosophy. In *Analytic Philosophy: An Interpretive History*, ed. Aaron Preston, 146–159. New York: Routledge.
Rorty, Richard. 1995. Response to Richard Bernstein. In *Rorty & Pragmatism*, ed. Herman J. Saatkamp Jr, 68–71. Nashville: Vanderbilt University Press.
Schlick, Moritz. 1936. Meaning and Verification. *Philosophical Review* 45 (4): 339–369.
Stadler, Friedrich. 2007. History of the Philosophy of Science. From *Wissenschaftslogik* (Logic of Science) to Philosophy of Science: Europe and America, 1930–1960. In *General Philosophy of Science: Focal Issues*, ed. Theo A.F. Kuipers, 557–658. Amsterdam: Elsevier.
Strassfeld, Jonathan. 2020. American Divide: The Making of 'Continental' Philosophy. *Modern Intellectual History* 17 (3): 833–866.
Tuboly, Adam Tamas. 2020. To the Icy Slopes in the Melting Pot: Forging Logical Empiricisms in the Context of American Pragmatisms. *HOPOS* 11 (1): 27–71.
Uebel, Thomas. 2015. American Pragmatism and the Vienna Circle: The Early Years. *Journal for the History of Analytical Philosophy* 3 (3): 1–35.
———. 2016. Pragmatisms and Logical Empiricisms: Response to Misak and Klein. *Journal for the History of Analytical Philosophy* 4 (5): 1–15.
Uebel, Thomas, and Christoph Limbeck-Lilienau. 2022. Introduction. In *The Routledge Handbook of Logical Empiricism*, 1–12. Abingdon: Routledge.
Van Eck, Nees Jan, and Ludo Waltman. 2010. Software Survey: VOSviewer, a Computer Program for Bibliometric Mapping. *Scientometrics* 84 (1): 523–538.
Verhaegh, Sander. forthcoming. The Analytic Turn in American Philosophy: An Institutional Perspective. Part II: Analytic vs. Continental Philosophy. *HOPOS*.
Verhaegh, Sander. 2020a. The American Reception of Logical Positivism: First Encounters (1929–1932). *HOPOS* 10 (1): 106–142.
———. 2020b. Coming to America: Carnap, Reichenbach, and the Great Intellectual Migration. Part I: Rudolf Carnap. *Journal for the History of Analytical Philosophy* 8 (11): 1–23.
———. 2023. Carnap and Quine: First Encounters (1932–1936). In *The Philosophical Project of Carnap and Quine*, ed. S. Morris, 11–31. Cambridge: Cambridge University Press.
———. 2024a. The Reception of Relativity in American Philosophy. *Philosophy of Science* 91 (2): 468–487.
———. 2024b. Logical Positivism: The History of a 'Caricature'. *Isis: A Journal of the History of Science Society* 115 (1): 46–64.
Waltman, Ludo, Nees Jan Van Eck, and Ed Noyons. 2010. A Unified Approach to Mapping and Clustering of Bibliometric Networks. *Journal of Informetrics* 4 (4): 629–635.

Chapter 8
Relativism and Sociology of Knowledge in Logical Empiricism: The Cases of Frank and Neurath

Martin Kusch

Abstract This paper is a response to two conflicting interpretations of the place of the sociology of knowledge and epistemic relativism in Logical Empiricism: the interpretations put forward by Thomas Uebel (*Philosophy of Science* 67:138–150, 2000) and David Bloor (*The Enigma of the Aerofoil: Rival Theories in Aerodynamics, 1909–1930*. Chicago and London: The University of Chicago Press, 2011). I agree with Bloor and Uebel that the sociology of knowledge was a central concern for Frank; and I am convinced by Uebel's arguments that the same is true for Neurath. I side with Bloor—against Uebel—in maintaining that Frank committed himself to relativism. I try to show the same for Neurath. Still, I disagree with important aspects of Bloor's rendering of Frank's relativism.

Keywords David Bloor · Thomas Uebel · Philip Frank · Otto Neurath · Sociology of knowledge · Relativism · Pluralism · Truth

8.1 Introduction

The expression 'relativism *in* logical empiricism' brings together two ideas that many philosophers regard as incompatible. After all, logical empiricism is widely thought of as a key chapter in twentieth-century "enlightened" or "scientific" philosophy, while relativism is often taken to be tantamount to subjectivism or irrationalism. The incompatibility of logical empiricism and relativism seems to be confirmed by Otto Neurath's book-length attack on a 'classic' of relativism, fittingly entitled "Anti-Spengler" (1921/1973).

A moment's reflection is enough, however, to appreciate that the situation is more complicated. For instance, Rudolf Carnap's "Principle of Tolerance"—a principle that contradicts Gottlob Frege's, Bertrand Russell's and Ludwig Wittgenstein's

M. Kusch (✉)
Department of Philosophy, University of Vienna, Vienna, Austria
e-mail: martin.kusch@univie.ac.at

G. Schiemer (ed.), *The Legacy of the Vienna Circle*, Vienna Circle Institute Yearbook 30, https://doi.org/10.1007/978-3-031-80568-4_8

insistence on the uniqueness or absoluteness of logic—has sometimes been flagged as a relativistic idea (Carnap 1937, §17). Or recall the famous *Protokollsatzdebatte*; in the context of that controversy, Moritz Schlick accuses Neurath of an allegiance to a "coherence theory of truth", and, in consequence, of a commitment to relativism (Schlick 1934).

Is it then, after all, appropriate to speak of 'relativism *in* logical empiricism'? Interestingly, for at least two influential interpreters of Carnap and Neurath the answer remains 'no'. Florian Steinberger (2016) argues that Carnap has the resources to make the choice between logical languages a fully rational, and thus anti-relativistic, process. And Thomas Uebel, the best-informed reader of the *Protokollsatzdebatte*, rejects Schlick's relativism charge by insisting that Neurath was no coherence theorist (2007: 8307).

In this paper, I shall argue for 'relativism *in* logical empiricism' from a different angle. I shall concentrate on two central figures, Philipp Frank and Neurath. I focus on Frank and Neurath for two reasons: On the one hand, both repeatedly positioned themselves explicitly as opponents of various kinds of *absolutism*. On the other hand, both were interested in the sociology of knowledge. This is noteworthy since the latter is one of the main fields of study where relativistic ideas have flourished since the nineteenth century. Moreover, Frank, Neurath, relativism, and the sociology of knowledge have featured in an important clash of interpretations between the anti-relativist Uebel and the self-proclaimed relativist and sociologist of knowledge, David Bloor. I am referring here primarily to Uebel's paper, "Logical Empiricism and the Sociology of Knowledge: The Case of Neurath and Frank" (2000), and a key chapter in Bloor's book *The Enigma of the Airfoil: Rival Theories in Aerodynamics, 1909–30* (2011).

Uebel and Bloor agree that Frank anticipates Bloor's "Strong Programme in the Sociology of Scientific Knowledge." Uebel claims much the same for Neurath. Given my purposes, it is crucial that Uebel and Bloor differ on the connection between the logical-empiricist version of the Strong Programme and relativism. Uebel insists that Neurath and Frank deny any such connection; Bloor is adamant that at least Frank clearly commits himself to relativism. I shall try to determine who is right, and right on what rendering of relativism.

I shall begin by explaining Uebel's main theses (Sect. 8.2), and Bloor's reply to Uebel's take on Frank (Sect. 8.3). In Sect. 8.4, I summarize my own position concerning Bloor's and Uebel's readings of Frank and Neurath. In Sect. 8.5, I provide evidence for my interpretation of Frank, and in Sect. 8.6 for my rendering of Neurath. Section 8.7 is my overall conclusion.

8.2 Uebel's Main Theses

Uebel seeks to establish that Frank and Neurath are committed to views in the vicinity of Bloor's "Strong Programme" in the "Sociology of Scientific Knowledge" (or "SSK" for short). Bloor formulates this programme as follows:

1. [The programme] would be causal, that is, concerned with the conditions which bring about belief or states of knowledge. Naturally there will be other types of causes apart from social ones which will cooperate in bringing about belief.
2. It would be impartial with respect to truth and falsity, rationality or irrationality, success or failure. Both sides of these dichotomies will require explanation.
3. It would be symmetrical in its style of explanation. The same types of cause would explain, say, true and false beliefs.
4. It would be reflexive. In principle its patterns of explanation would have to be applicable to sociology itself. (Bloor 1991, 7)

A first piece of evidence for Neurath's leaning towards Bloor's Strong Programme is the following. Neurath insists that natural and social scientists act always already within the wider society and seek to influence it. Neurath expresses this idea by writing that "[t]here is no extraterritoriality for sociologists, ... or other scientists" (1944a, 46), or that "trends in scientific research are never socially neutral ..." (Neurath 1931a/1973, 403) (Uebel 2000, 140).

Moreover, Uebel's Neurath accepts "the argument from underdetermination to external influences on theory choice ..." (Uebel 2000, 143) Neurath puts the point in this way: "... due to the insufficiency of our knowledge of the available data, our predictions are multiply ambiguous. It is resolution that must decide. And this is often historically determined by the traditional forms of cognitive co-operation." (Neurath 1930a/1981, 352) Uebel's Neurath also defends a version of the Bloorian *Reflexivity* principle. Marxism, Neurath holds, helps us identify and analyse ideologies, but Marxism itself too is an ideology (Uebel 2000, 146).

Similar statements can also be found in Frank. Frank claims, for example, that so-called "'scientific' criteria" of theory choice "in the narrower sense—agreement with observations and logical consistency"—are "often" not enough. In such cases, other elements must come into play: considerations of the "purpose of a theory ... simplicity and elegance, agreement with common sense, fitness to support desirable human conduct etc." (Frank 1957/2004, 6993) Finally, according to Uebel, the already cited passages also indicate that Neurath and Frank accept Bloor's *Symmetry* principle, as well as *Causality* and *Impartiality* (Uebel 2000, 145).

Uebel's paper has two main theses. The first is that Frank and Neurath are envisaging a form of sociology of knowledge that Bloor, half a century later, sums up in his "Strong Programme." This is the thesis supported by the passages quoted above. Uebel's second central thesis concerns an alleged *difference* between the two logical empiricists on the one side, and Bloor on the other side: Frank and Neurath dismiss any thought of a link between the Strong Programme and relativism, a link that Bloor sees as essential. Uebel's evidence is that Frank and Neurath reject the idea that "there is nothing left to rationality but 'mob rule'," and that both emphasize a "distinction between descriptive and normative inquiries." (Uebel 2000, 148).

The last two sentences tell us something important about Uebel's understanding of relativism. Uebel thinks that to be a relativist is to replace rationality with "mob rule" and to refuse consideration of normative questions. It is easy to see the connection between these two elements: if there is no rationality over and above mob rule, than might is right. As Uebel has it, what sociologists of knowledge can learn from the Vienna Circle is precisely how to engage in the sociology of knowledge without a commitment to relativism as mob rule.

8.3 Bloor's Main Theses

Bloor's reply to Uebel comes 11 years after "Logical Empiricism and the Sociology of Knowledge" and confines itself to a discussion of Uebel's reading of Frank. Bloor concurs with Uebel's claim that Frank anticipated the Strong Programme. (Bloor 2011, 428) Still, Bloor offers further 'smoking guns' for this rendering. He is particularly impressed by passages in Frank's paper "On the Variety of Reasons for the Acceptance of Scientific Theories" (1954), and Frank's *Relativity: A Richer Truth?* (1951a), published in German as *Wahrheit—Relativ oder Absolut?* (1952).

In the 1954-paper, Frank maintains that the theory-choice criteria of "simplicity" and "explanation of observed facts" lead to determinate decisions only if the scientists in question have already answered several prior questions: Which of the observed facts are to be explained? What type of simplicity should be used? And what should we do when two criteria conflict? In order to answer these prior questions, Bloor's Frank suggests, it needs "the social conditions that produce the conditioned reflexes of the policy-makers" (Frank 1954, 144; Bloor 2011, 428).

Bloor finds in Frank's 1954-paper also reasons to disagree with Uebel. The main disagreement turns on the issue of relativism. As Bloor has it, Frank does unmistakably adopt relativistic theses. Frank claims that since there is "no absolute, final, or perfect compromise" between the various criteria for theory choice, no existing compromise can be regarded as "the truth." Rather than think of a scientific theory as true or false, Frank urged his readers to understand it as "a tool that produces other tools according to a practical scheme." (Frank 1954, 144) This is relativism as Bloor understands it (Bloor 2011, 428).

Furthermore, Bloor laments the fact that Uebel takes little notice of Frank's *Relativity: A Richer Truth?* Bloor takes the upshot of the book to be that scientists are *cognitive* relativists, and that all of us—scientists and non-scientists alike—are *moral* relativists. The work allegedly establishes that "absolute truth" is strictly speaking meaningless, never mind that the expression is "woven into the fabric of life." Bloor's Frank also adheres to the view that scientific knowledge is relative "to whatever causes determine it." Causation itself is, Bloor believes, a relativistic idea, and he credits Hume with first having worked out the connection between causation and relativism. (Bloor 2011, 430)

Needless to say, in placing Frank squarely in the relativistic camp, Bloor is not denouncing Frank as a champion of "mob rule." Bloorian relativism consists of the ideas that scientific knowledge is the product of contingent (in part social) causes; that scientific theories are tools in the service of local and contingent goals; and that no sense can be made of the notion of "absolute truth." (Cf. Kusch 2020)

8.4 Bloor or Uebel: Who Is Right?

Who then is right about Frank's stance *vis-à-vis* relativism—Uebel or Bloor? There is a way to avoid having to choose: perhaps Frank *is* a relativist in Bloor's sense of the term; and *is not* a relativist in Uebel's sense. That is, perhaps Frank's rejects mob rule while still maintaining that "absolute truth" is s meaningless expression. This 'Solomonic judgement' may well be accurate but it leaves us with a further question: What rendering of relativism should we adopt? Since this paper is primarily historical, I propose that we follow our historical actors; that is, that we seek to determine Frank's and Neurath's own understanding of relativism. Let there be no false suspense: In what follows I shall try to show that Frank's (and Neurath's) notion of relativism is much closer to Bloor's than to Uebel's conception. This is the first thesis of the remainder of this paper.

My second thesis is that Bloor and Uebel are spot-on in suggesting that Frank and Neurath anticipated key elements of Bloorian SSK. And yet, both Bloor and Uebel seem to me to overstate the similarity between the three versions of the sociology of knowledge. My third thesis targets another excessive assimilation: I disagree with Bloor regarding the central message of Frank's *Relativity: A Richer Truth?* The book's account of relativism differs fundamentally from the relativism of the 1954-paper, or the version of relativism Bloor himself argues for.

Fourth, as far as Neurath is concerned, I maintain that Uebel pays too little attention to the numerous places where Neurath explicitly attacks a wide variety of absolutisms, and explicitly defends relativism and pluralism. Fifth, and finally, Uebel thinks that Frank and Neurath teach us how to do SSK without relativism. I believe this claim to be false—at least when "relativism" is understood in the ways in which Frank and Neurath themselves (implicitly) defined the term.

8.5 Frank on Relativism and the Sociology of Knowledge

I begin with Frank. One valuable and central source in what follows is a text Uebel and Bloor were unable to use: Frank's *The Humanistic Background of Science*, written in the late 1950s, and published for the first time in 2021 (Frank 2021). I shall discuss Frank's views on relativism and SSK in four steps. I begin with the relationship between science and philosophy.

Bloor has no time for philosophy, famously calling SSK—with Wittgensteinian overtones—an "heir to the subject that used to be called 'philosophy'" (1991, 52), and complaining elsewhere that "to ask questions of the sort which philosophers address to themselves is usually to paralyze the mind" (1983, 182). Accordingly, Bloor rebukes philosophers like Michael Friedman (1998) or Elizabeth Nemeth (2003) who task philosophy with "taking reflective responsibility" for the sciences. Bloor assumes, without much argument, that Frank would agree with him.

This is wrong: Frank does not share Bloor's animus towards philosophy. For instance, Frank notes that "fundamental changes in science have always been accompanied by deeper digging toward the philosophical foundations." (1957/2004, 73) Frank gives philosophy a crucial role in the development of the sciences: philosophy is inseparable from the sciences and essential for their integration and unification. Philosophy suggests a "common language" for the sciences and protects them from attacks coming from metaphysicians and common sense. Philosophy also reminds scientists of the role of social-political values in their work (2021: 2413, 4588). Frank calls the philosophy needed here "proud, active positivism" (1951b, 30). Frank's position is thus not exclusively sociological. And Nemeth is right to describe Frank's philosophy—using Friedman's term—as aiming to "take reflective responsibility" for the sciences (Friedman 1998, Nemeth 2003).

How then do Frank's views fit with Bloor's Strong Programme more generally? Here there is a lot of common ground between Uebel, Bloor and myself. I can add to their case only marginally. Underdetermination of theory choice by data and by other theories is indeed a central topic for Frank. And there is no denying that he recognizes the role of values, interests and political ideas in reducing the underdetermination in practice. Moreover, Frank rejects the view that "in any period there is [but] one theory that is 'accepted by scientists'." He continues:

> This is certainly not the case if we understand by 'theories' a system of statements of the highest level of abstraction. Even in the field of mathematical physics and astronomy, we don't know whether the assumption of a finite universe or of an infinite universe checks better with observed facts. (1951b, 24)

Moreover, it is also noteworthy that Frank's criticism of Mannheim has much in common with Bloor's critique. Frank laments Mannheim's restricting the sociology of knowledge to the humanities and the social sciences, and he ridicules the idea of the politically neutral "freely floating intelligentsia." (2021: 3263) In the same passage, Frank signals some sympathy for Bloor's *Symmetry* principle by writing that "social factors [are] involved in … building up 'true theories'." In other words, "social factors" are not only present when false theories prevail.

So far, so good for Bloor and Uebel. Still, there are also passages in Frank that do not fit neatly into the two scholars' interpretations. First and foremost, I doubt that Frank is fully committed to Bloorian *Symmetry*. Bloor's *Symmetry* states that true and false theories have "the same types of cause", and that, amongst these types of causes, invariably are social causes. Frank's position does not go quite so far. While he is adamant that both true and false theories have social causes (amongst other causes), he does not say anything on whether or not the social causes of true and false theories are of the same types.

Moreover, both Bloor and Uebel overlook that Frank's rudimentary sociological-political focus on science is often concerned with justification and critique. (This is of course connected to the already mention idea of "taking reflective responsibility" for the sciences.) That is to say, Frank wants the social sciences and philosophy to investigate which social purposes natural scientific theories are meant to serve, whether these purposes are legitimate, and whether the theories in question are *in fact* fulfilling their social-political promises.

Frank puts the point as follows. Comte organized the sciences hierarchically in the order: "mathematics, astronomy, physics, chemistry, biology, and sociology. ... The main point was that every science presupposes the sciences that were prior in this series but not the posterior ones." (2021: 139) Comte's hierarchy has been refuted, Frank submits, by what we have learnt know from the history of science in general, and the social history of science in particular. We must therefore follow Émile Littré and replace the Comtean linear hierarchy with a circle: "mathematics, astronomy, physics, chemistry, biology, sociology, and then again mathematics, astronomy, etc." (2021: 140) There is thus a sense in which mathematics, astronomy, physics, chemistry, and biology *presuppose* sociology.

This is no longer Bloor's Strong Programme; instead, it is reminiscent of certain positions in feminist epistemology. I am thinking here, first and foremost, of Sandra Harding's claim that: "We could say that the natural sciences should be considered to be embedded in the social sciences because everything scientists do or think is part of the social world." (1991, 99) Harding's call for "strong objectivity" aims for the kind of social-political reflection that Frank was gesturing towards.

I now turn to Frank's relativism. *The Humanistic Background of Science* offers some evidence that Frank was happy to speak of theories as "relatively true", and of entities as "existing relatively":

> ... we can ascribe the geocentric or heliocentric system only "relative truth". The heliocentric theory is relatively true if we insist on simplicity, the geocentric theory is "true relative to" the requirement of agreement with common sense or agreement with Aristotelian and Thomistic philosophy. (2021: 4473)

> ... whether, for example, this piece of wood we call a "table" is "in reality" an agglomeration of electrons and nucleons between which there is empty space, ... we can only answer: it depends on what results we want to derive from the statements in which the word "table" occurs. (2021: 2880)

These passages support Bloor's reading of Frank as a relativist, and they make trouble for Uebel's account. Still, there are problems also with Bloor's interpretation, and especially in *Relativity: A Richer Truth?*, the book that Bloor gives special prominence in Frank's relativism. The main source of my unease is Frank's speaking of "relativism" as a "method" or "attitude" for making scientific language richer and thus more precise (1951a, 20). Take the sentence:

> (a) "My head is above my feet ..."
>
> ... said by me here and now. Considering Smith, currently standing on the other side of the globe, I am inclined to say:
>
> (b) Smith's head is below his feet.
>
> Smith is likely to disagree and to insist on her right to utter (a) as well. Scientific progress, Frank suggests, has enabled us to get rid of this kind of disagreement. The key was to relativize "above" and "below" to "the field of gravity". My (a) should thus be taken to mean:
>
> (c) "My head is above my feet relative to the field of gravity at my place."

And Smith can say the same relative to her field of gravity. (Frank 1951a, 23) Thus amended, Frank writes, "[b]oth answers are 'absolutely'—or better, dependably true." (1951a, 24) Frank goes on to argue that scientific progress involves the construction of an ever "richer language" (with ever more such qualifications). Using such richer language, "our assertions remain as definite, as 'objective,' as 'absolute,' as the statements of the simpler type which we used" before. (1951a, 25) With the appropriate qualifications, we "come one step nearer to what one may call the 'plain truth about the universe'." (1951a, 30) That does not sound like relativism as ordinarily understood, or as Bloor himself understands it.

Let me sum up. Uebel and Bloor are right to find in Frank's work anticipations of SSK. And yet, there are also other elements that do not fit well with classifying Frank as a proto-Bloorian: Frank's take on *Symmetry*, his view of philosophy, and the normative role of the social sciences *vis-à-vis* the natural sciences. Bloor is right to stress—pace Uebel—that Frank was happy to call himself a "relativist". When Frank speaks of theories and entities as relative to background theories, his rendering of relativism seems close to Bloor's, not to Uebel's. But *Relativity: A Richer Truth?* goes in a different direction. Terms are relativized so that we can get closer to the truth. Finally, Frank does not aim for a Strong Programme without relativism. In *The Humanistic Background of Science,* he accepts the (Bloorian) relativism that goes with social-historical studies of science.

8.6 Neurath on Relativism and the Sociology of Knowledge

Bloor confines his discussion to Frank; thus, in this section I 'only' have to contend with Uebel, the doyen of Neurath-studies. Uebel has repeatedly insisted on Neurath's opposition to relativism (2000, 2007, Cartwright et al. 1996). Uebel claims that Neurath proves his anti-relativism once and for all in his 1921-book *Anti-Spengler*. And Uebel rejects Schlick's accusation that Neurath's so-called "coherence theory" committed him to relativism. I beg to differ. Neurath is a self-proclaimed relativist, and his relativism is recognizable as such also from today's perspective.

Anti-Spengler must be addressed first. *Pace* Uebel, I read the book as being *ambivalent* about relativism: it does not refute cultural relativism *per se*; and it aims to establish that Spengler does not *make a good case* for cultural relativism. Spengler fails to show that shared beliefs and values across cultures are impossible. But there are ways in which he could have demonstrate why at least a limited form of incommensurability holds. First, and foremost, Neurath suggests, Spengler could have used Duhem, Poincaré, and Dilthey, and generalized from the underdetermination of theories by observations, to the underdetermination of worldviews by data. (Neurath 1921/1973, 206)

Some of the *pro-* and *contra*-arguments of *Anti-Spengler* resonate in later writings. For instance, in 1922 Neurath expresses an interest in Hermann Graf Keyserling's cultural-relativist *Reisetagebuch eines Philosophen* (1919), and he calls for a "*comparative Voelkerpsychologie*" to follow up on Keyserling's work

(Neurath 1922/1981, 206). Some of the criticisms of Spengler are later used—*mutatis mutandis*—against others. For example, Neurath argues, contra Lucien Lévy-Bruhl, that humans who believe in magic do not have a different logic from Europeans: it is the content, not the form of thought that differs. (Neurath 1931a/1973, 321) And in still later texts (from 1941 and 1944), Neurath urges that all languages share a common core. But here too relativistic motifs are present. Neurath proclaims that in addition to the common, inter-translatable, core there are also areas where translation is impossible: "speculative metaphysics" (1944b/1981, 932), ethics or politics. (1941/1981, 223)

Turning to Neurath's ideas on the social sciences and the sociology of knowledge, Uebel is right to identify traces of Bloorian *Symmetry*, *Causality*, *Neutrality*, and *Reflexivity* in Neurath. But just like Frank, so also Neurath nowhere commits himself to the view that the *very same types of causes* (social or other) explain true and false scientific beliefs. Moreover, Uebel misses that, for Neurath, the sociology of knowledge is linked to relativism. For instance, Neurath rejects as unworkable Mannheim's attempt to avoid relativism by invoking the "freely-floating intelligentsia" with its allegedly neutral and objective standpoint. Like Frank and Bloor, so also Neurath does not follow Mannheim in exempting the natural sciences from the domain of the sociology of knowledge; as Neurath had it, even strictly natural-scientific forms of analysis can end up in fundamental and undecidable disagreements, disagreements that cannot be resolved merely by more data and more theory alone. (Neurath 1930a/1981, 352)

Neurath's main project regarding the social sciences was not, however, the development of a sociology of knowledge. The main project was to cleanse the social sciences of the metaphysical tendencies and "absolutist formulations" of Weber, Mannheim or Scheler. (1930a/1981, 352–3; 1931a/1973: 349; 1936g/1981, 773–4) A properly understood sociology, claims Neurath, encourages an intellectual "modesty" that resists "all absolutist" positions. (1936a/1981, 769) Neurath holds that the social sciences are essentially "pluralist": for instance, the biography of Galileo can be written in many different ways. (1944b/1981, 940) Neurath demands even of Marxism a recognition of this pluralism. (1931a/1973, 386–89; 1944b/1981, 974)

As Neurath puts it in letters to Carnap, "pluralism" is the idea that in science and political life "we start with many divergent statements and remain with divergent statements forever, as it were." (Letter from September 9th, 1943; Damböck et al. 2024) Duhem, Poincaré, Frank and Neurath are genuine pluralists; Hempel, Carnap are pluralists "to a certain extent" only; Tarski and Popper "represent a non-pluralist tendency." (Letter from April 1st, 1944; Damböck et al. 2024) The pluralist insight, most easily gained in sociology, must also become central in our understanding of the natural sciences. In this vein, Neurath refers to Duhem's and Poincare's ideas about the underdetermination of theory by observation, confirmational holism or conventionalism. (1938/1981: 870)

The thought of radicalizing Duhem's and Poincaré's ideas had been central for Neurath since the 1920s. "Radicalizing" the two French philosopher-scientists means to go beyond the underdetermination of theory by protocol sentences, and to add the underdetermination of protocol sentences, and the underdetermination of the priority between theories and protocol sentences. Neurath expresses the resulting confirmational holism by saying: "Science with all its parts is always up for discussion." (1937a/1981, 793) The issue is related to the dispute with Schlick. Neurath allows for a plurality of consistent systems of sentences between which only "extra-logical moments" can decide. For Schlick this is "a logical impossibility." As examples of such "extra-logical moments", Neurath alludes to the "primacy of life" and the need for coordinated action. (Neurath 1934/1983, 107–110) This position is not far from what Bloor calls "relativism".

We get a still better grasp of Neurath's overall position by turning to the topics of epistemology and "pseudorationalism." For Neurath, epistemology is an absolutist project, and a form of pseudorationalism. In 1931 Neurath writes: "… there can be no 'theory of knowledge,' at least not in the traditional form. It could only consist of defence actions moves against metaphysics …" (1931b/1983, 67) Part of the animus here is against Wittgenstein's and Schlick's "absolutist" thought that our knowledge can be analysed into elementary sentences that in turn can be immediately verified. (Letter by Neurath to Carnap, March 30th, 1935; Damböck et al. 2024) Popper's position does fare no better. It implies an "absolutism of certainty," an "absolutism of precision" (1936a/1983: 147), and an "absolutism of falsification" (1935b/1983: 131). Neurath's alternative is summed up in the sentence: "We constantly have to change the totality of our sentences, including the principles of confirmation." (1936e: 766) "Falsification" is thus only one of many possible principles for theory choice.

Neurath first targets pseudorationalism as early as 1913. Pseudorationalism denies the role of "judgement" (beyond rules) in the practical and theoretical domains. (Neurath 1913/1983) This thought remains central throughout Neurath's life. E.g. in 1937 he argues that judgement is crucial in the choice of theories and protocol sentences. (1937b/1983) And in Anti-Spengler, Neurath criticizes Spengler's followers for mistaking their "decision" to accept Spengler's claim asd a proof. (1921/1973: 160)

We can also connect Neurath's "encyclopedism" and "physicalism" to relativism. From the late 1930s, Neurath worries that the project of *Einheitswissenschaft* contradicts his relativism: he fears that *Einheitswissenschaft* becomes a single "hierarchical system," "absolute and pyramidistic" (1937b/1983: 203–4). Talk of "encyclopedias" as well as the "mosaic" of the sciences is meant to correct these misleading images. "Encyclopaedia" variously stands for: "a preliminary assemblage of knowledge," a provisional collection of knowledge; the most perfect form of summarising the sciences at a given time; an actual or possible structure connecting the sciences; or the philosophical project of creating such a structure. (1936b/1983, 146, 157; 1936d/1983, 137–8; 1938/1981, 855) In all these senses, an encyclopaedia is a product of history, and it cannot be meaningfully contrasted with an "'extra-historical' ideal." (1936b/1983, 157) In other words, the theory of encyclopedias has to "acknowledge the demand of relativism" and avoid "the traditional absolutistic terminology that allows reference to 'the real world', the 'ideal totality of statements' and other similar things." (1936b/1983, 147)

The last quotes invite further comments on the topics of world and truth. To fully understand Neurath's position here we must return to Frank, and especially to his 1932-book *Das Kausalgesetz und seine Grenzen*. The key chapter for our concerns is Chapter X, entitled "Of the So-called True World" (1932, 252–286). Frank targets the metaphysical assumption that our theories and our measurements get ever closer to the way the world is. (1932, 296) This metaphysical thought, Frank argues, implicitly commits one to the theological idea of a "higher intelligence" whose theory about the world we approximate. (1932, 293–4) Frank claims that scientific progress consists in ever "better ... orderings of experiences," where the choice between different senses of "better" is "arbitrary." (1932, 294) The precision of our predictions of experiences might increase; but this gives us no licence to commit to the theoretical entities of our theories. In other words, Frank urges his readers not to conflate the convergence of the numerical values calculated using our theories with the convergence of these theories themselves (1932, 297). For instance, the theoretical entities and assumptions of the astronomical theories of Copernicus, Kepler, Newton and Einstein are very different. Frank also mentions that leading quantum physicists counsel caution concerning expressions like "the real world" (over and above our experiences). (1932, 304)

Neurath credits Frank with being the first in the Vienna Circle who challenged talk of "*the* reality" (Letter by Neurath to Carnap, May 5th, 1936; Damböck et al. 2024) Neurath writes an enthusiastic review of Frank's book (Neurath 1932/1981). Neurath agrees with Frank that talk of perfect knowledge of reality as such, even as a limit, is a theological element. (1930b/1983, 32) And thus no sense can be made of the history of science as an "asymptotic approximation to the 'truth'." (1936c/1981, 701) Neurath wholeheartedly agrees with Frank's antirealism and relativism. But Neurath also goes beyond Frank. Neurath replaces Frank's talk of experiences with talk of sentences. Focusing on experiences inclines one towards individualism or even solipsism. Taking one's starting point from "the social implications of language" facilitates work on "the history and sociology of the sciences" and undermines all forms of absolutism. (1941/1981, 918)

Neurath draws several conclusions from his revised antirealist position. First, he rejects the correspondence theory of truth. The latter is an "absolutist conception of truth" and undermined by Frank's arguments. (1935a/1981: 665) Neurath also labels Tarski's "Convention T" "absolutist." In ...

'p' is true if, and only if, p

... Neurath maintains, the consequent is treated as something "absolute" and is "ontological talk." (1944b/1981, 938; cf. letter by Neurath to Carnap, December 23rd, 1935, Damböck et al. 2024) Tarski's theory does not capture the "emotional emphasis" of "philosophical, religious 'truths'." It only works for formal languages. (Letters from Neurath to Carnap, December 8th and 23rd, 1935, Damböck et al. 2024)

Second, Neurath offers a number of proposals on how to deal with truth and related concepts: Stop using "true," "verification," "progress," "reality" altogether. (1936f/1983: 169; 1937c/1983: 180). Engage in 'conceptual engineering' and offer a new understanding of truth along roughly coherentist lines (1936b/1983: 135). Or replace talk of truth with talk of what is 'accepted by us at a given time and place'. This avoids all "absolute uses" of "true." (1944b/1981: 938–9)

Third, in suggesting, in 1936, a coherentist rendering of truth as a form of conceptual engineering, Neurath is of course referring to his position in the *Protokollsatzdebatte*. It seems fair to say that Neurath's rendering of truth and justification in that debate has coherentist elements. And he happily owns the relativistic implications. Here is a key quote:

> If a statement is made, it is to be confronted with the totality of existing statements. If it agrees with them, it is joined to them; if it does not agree, it is called "untrue" and rejected; or the existing complex of statements of science is modified so that the new statement can be incorporated; the latter decision is mostly taken with hesitation. *There can be no other concept of 'truth' for science.* (1931b/1983, 53)

Still, Neurath denies that his is a coherence theory of truth. And he does so on varying grounds. For instance, Schlick (1910) claims that coherence theory and idealism (in Windelband, Joachim and Bradley) go hand in hand. But then Neurath cannot be a coherence theorist since he rejects idealism. (Letter by Neurath to Carnap, May 22nd, 1934) Another consideration is that coherence theorists give no special weight to protocol sentences; but Neurath does. (1934/1983, 110)

Uebel (2007, 8307) defends Neurath against Schlick's relativism-charge by insisting that Neurath is no coherence theorist. But Uebel says nothing about Neurath's owning relativism. Neurath criticises Schlick for thinking that to call a view "relativistic" is to discredit it—a bit like calling a view "contradictory." But this is just wrong, Neurath writes. (1934/1983, 103) Moreover, Neurath regards Schlick's attempts to find propositions that are an "absolutely secure foundation of knowledge" was metaphysical and idealist. (1934/1983, 101–3) Schlick is wrong to also claim that even relativists aim to find "the one true world." Frank has shown that this expression is meaningless. (1934/1983, 109)

I shall now draw some conclusions concerning Neurath. Uebel is right is this sense: there are relativist elements Neurath rejects. These elements are: radical incommensurability across cultures in all domains; the thought that 'true' and 'false' are relative to Spenglerian cultures; or the notion that different cultures have different 'logics' (cf. Lévy-Bruhl's '*mentalité primitive*'). But there is also a long list of items Neurath calls "absolutist" and rejected: talk of 'reality as such', or of 'true reality'; truth as correspondence; Mannheim's "free-floating intelligentsia"; Schlick's and Wittgenstein's foundationalism; scientific realism; fixed standards or norms in epistemology or politics; pseudorationalism; pyramidism, monism, and unrestricted commensurability. And there are many relativist tropes that Neurath endorses: the idea of deep cultural differences; various forms of underdetermination: of worldviews, of theories, of the priority between theory and Protokollsätze; social values and interests are elements needed to overcome underdetermination; pluralism in the sciences and politics; irresolvable disagreements; truth as coherence or consensus; truth as a concept to be re-engineered; truth as an obsolete concept; incommensurability in certain areas: for instance, in metaphysics, ethics and politics; or the historicity of encyclopedias.

8.7 Conclusion

Time has come to zoom out from the details and highlight a few more general lessons. First, it is wrong to think that the Vienna Circle was committed to an anti-relativist philosophy or stance. Second, some key figures of the Vienna Circle saw the need for, and potential of, a sociology of scientific knowledge. Third, at one stage Neurath claimed that Frank, Hans Hahn and himself had been "philosophiefrei" from early on. (1933/1981, 588) But this is not Bloor's total rejection of philosophy. Both Neurath and Frank continue to ask normative questions and seek to influence the course of science and society. Fourth, Uebel is right that Frank and Neurath were not relativists if by 'relativism' we mean the view that rationality boils down to mob rule and a dismissal of all normative inquiry. But this was not Frank's and Neurath's notion of relativism. Their notion is closer to Bloor's.

Fifth, Neurath's and Frank's relativism is even more expansive than Bloor's: further elements are opposition to scientific realism, anti-foundationalism in epistemology, scientific pluralism, pragmatism, operationalism, particularism, historicism, coherence and consensus theories of truth and justification, and an acceptance of incommensurability in at least some areas of discourse. Sixth, we should not immediately denounce this as a confusion. At least as historians of philosophy we should map out the changing renderings of relativism and try to understand the development.

And finally, seventh, understanding relativism as something like the more expansive position of Frank and Neurath, it not without interest in today's philosophical world either. In my recent little book on "relativism in the philosophy of science" (Kusch 2020), I too sketch such position in conversation with Bas van Fraassen and Bloor. I was not aware when writing the book that I was going over a field first ploughed by Frank and Neurath. *Mea culpa.*

References

Bloor, David. 1983. *Wittgenstein: A Social Theory of Knowledge*. New York: Columbia University Press.

———. 1991. *Knowledge and Social Imagery*. 2nd ed. Chicago, IL: The University of Chicago Press.

———. 2011. *The Enigma of the Aerofoil: Rival Theories in Aerodynamics, 1909–1930*. Chicago and London: The University of Chicago Press.

Carnap, Rudolf. 1937. *The Logical Syntax of Language*. London: Routledge.

Cartwright, Nancy, Jordi Cat, Lola Fleck, and Thomas E. Uebel. 1996. *Otto Neurath: Philosophy between Science and Politics*. New York: Cambridge University Press.

Damböck, Christian, Johannes Friedl, and Ulf Höfer, eds. 2024. *Rudolf Carnap und Otto Neurath: Briefwechsel*. Hamburg: Meiner Verlag.
Frank, Philip. 1932. *Das Kausalgesetz und seine Grenzen*. Vienna: Springer.
———. 1951a. *Relativity: A Richer Truth*. London: Jonathan Cape.
———. 1951b. The Logical and Sociological Aspects of Science. *Proceedings of the American Academy of Arts and Sciences* 1: 16–30.
———. 1952. *Wahrheit—Relativ oder absolut?* Zurich: Pan-Verlag.
———. 1954. On the Variety of Reasons for the Acceptance of Scientific Theories. *Scientific Monthly* 79: 139–145.
———. 1957/2004. *Philosophy of Science: The Link between Science and Philosophy*. Mineola, NY: Dover Publications. Kindle edition.
———. 2021. In *The Humanistic Background of Science*, ed. George A. Reisch and Adam Tamas Tuboly. Albany, NY: State University of New York Press.
Friedman, Michael. 1998. On the Sociology of Scientific Knowledge and Its Philosophical Agenda. *Studies in the History and Philosophy of Science* 29: 239–271.
Harding, Sandra. 1991. *Whose Science? Whose Knowledge? Thinking from Women's Lives*. Ithaca, NY: Cornell University Press.
Keyserling, Graf Hermann. 1919. *Das Reisetagebuch eines Philosophen. Zwei Bände*. Darmstadt: Reichl.
Kusch, Martin. 2020. *Relativism in the Philosophy of Science*. Cambridge: Cambridge University Press.
Nemeth, Elisabeth. 2003. Philosophy of Science and Democracy: Some Reflections on Philipp Frank's *Relativity: A Richer Truth*. In *Philosophy of Science and Politics*, ed. Michael Heidelberger and Friedrich Stadler, 119–138. Vienna: Springer.
Neurath, Otto. 1913/1983. *The Lost Wanderers of Descartes and the Auxiliary Motive*. Repr. in Neurath 1983, 1–12.
———. 1921/1973. *Anti-Spengler*. Repr. in Neurath 1973, 158–213.
———. 1922/1981. *Weltsozialismus*. Repr. in Neurath, 1981, 203–208.
———. 1930a/1981. *Bürgerlicher Marxismus*. Repr. in Neurath, 1981, 349–356.
———. 1930b/1983. *Ways of the Scientific World-Conception*. Repr. in Neurath 1983, 32–47.
———. 1931a/1973. *Empirical Sociology: The Scientific Content to History and Political Economy*. Repr. in Neurath 1973, 319–421.
———. 1931b/1983. *Physicalism*. Repr. in Neurath 1983,52–57.
———. 1932/1981. *Die 'Philosophie' im Kampf gegen den Fortschritt der Wissenschaft*. Repr. in Neurath, 1981, 571–576.
———. 1933/1981. *Einheitswissenschaft und Pychologie*. Repr. in Neurath, 1981, 587–610.
———. 1934/1983. *Radical Physicalism and the 'Real World'*. Repr. in Neurath 1983, 100–114.
———. 1935a/1981. *Erster Internationaler Kongress für Einheit der Wissenschaft in Paris 1935*. Repr. in Neurath, 1981, 649–672.
———. 1935b/1983. *Pseudorationalism as Falsification*. Repr. in Neurath 1983, 121–131.
———. 1936a/1981. *Ansprache in der Eröffnung*. Repr. in Neurath, 1981, 769–770.
———. 1936b/1983. *Encyclopedia as 'Model'*. Repr. in Neurath 1983, 145–158.
———. 1936c/1981. *Die Entwicklung des Wiener Kreises und die Zukunft des Logischen Empirismus*. Repr. in Neurath, 1981, 673–702.
———. 1936d/1983. *Individual Sciences, Unified Science, Pseudorationalism*. Repr. in Neurath 1983, 132–138.
———. 1936e/1981. *Drei Diskussionsbeiträge zum 8. Internationalen Kongress für Philosophie*. Repr. in Neurath, 1981, 765–768.
———. 1936f/1983. *Physicalism and the Investigation of Knowledge*. Repr. in Neurath 1983, 159–171.
———. 1936g/1981. *Soziologische Prognosen*. Repr. in Neurath, 1981, 771–776.
———. 1937a/1981. *Prognosen und Terminologie in Physik, Biologie, Soziologie*. Repr. in Neurath, 1981, 787–794.

———. 1937b/1983. *The Departmentalization of Unified Science*. Repr. in Neurath 1983, 200–205.
———. 1937c/1983. *Unfied Science and its Encyclopedia*. Repr. in Neurath 1983, 172–182.
———. 1938/1981. *Die neue Enzyklopädie*. Repr. in Neurath, 1981, 863–872.
———. 1941/1981. *Die Gefahr sorgloser Terminologie*. Repr. in Neurath, 1981, 919–924.
———. 1944a. *Foundations of the Social Sciences*. Chicago, IL: University of Chicago Press.
———. 1944b/1981. *Grundfragen der Sozialwissenschaften*. Repr. in Neurath, 1981, 925–978.
———. 1973. *Empiricism and Sociology*, ed. Marie Neurath and Robert S. Cohen. Dordrecht: Reidel.
———. 1981. *Gesammelte philosophische und methodologische Schriften*, ed. Rudolf Haller and Rainer Hutte. Vienna: Hölder-Pichler-Tempsky.
———. 1983. *Philosophical Papers 1913–1946*, ed. Robert S. Cohen and Marie Neurath. Dordrecht: Reidel.
Schlick, Moritz. 1910. Das Wesen der Wahrheit nach der modernen Logik. *Vierteljahrsschrift für wissenschaftliche Philosophie und Soziologie* 34: 386–477.
———. 1934. Über das Fundament der Erkenntnis. *Erkenntnis* 4: 79–99.
Steinberger, Florian. 2016. How Tolerant Can You Be? Carnap on Rationality. *Philosophy and Phenomenological Research* 92: 645–668.
Uebel, Thomas E. 2000. Logical Empiricism and the Sociology of Knowledge. *Philosophy of Science* 67: 138–150.
———. 2007. *Empiricism at Crossroads: The Vienna Circle's Protocol-Sentence Debate*. Chicago and La Salle, IL: Open Court. Kindle Edition.

Chapter 9
Scientific Pluralism as a Branch of Millenial Popperianism?

Alexander Linsbichler, Sophie Juliane Veigl, and Benjamin Smart

Abstract Of all the positions that have amassed intellectual traction within the philosophy of science in the previous decades, it is scientific pluralism that is most often pitted against Karl Popper's image of the sciences. In fact, many card-carrying scientific pluralists explicitly distance their philosophies from Popper's critical rationalism. Particularly, they consider pluralism about scientific theories or explanations to be in opposition to falsificationism. In this paper we shall critically re-examine this claim and the relation of Popper's philosophy to millennial scientific pluralism. Following Kellert and colleagues, we distinguish two flavours of scientific pluralism: radical and modest. For both flavours, we examine congruities between pluralism and Popper's work. While we acknowledge crucial differences, we argue that these tend to be overstated, and that there is much room for fruitful exchange. More specifically, we highlight that Popper endorses and embraces plurality with respect to many types of scientifically relevant entities including theories, models, and practices; and we qualify Popper's purported fixation on truth as the only objective of science.

A. Linsbichler (✉)
Institute of Philosophy and Scientific Method, Johannes Kepler University Linz, Linz, Austria

Department of Philosophy and Department of Economics, University of Vienna, Vienna, Austria

Department of International Economics, Vienna School of International Studies (Dipomatische Akademie), Vienna, Austria
e-mail: alexander.linsbichler@jku.at

S. J. Veigl
Department of Philosophy, University of Vienna, Vienna, Austria
e-mail: sophie.juliane.veigl@univie.ac.at

B. Smart
Centre for the Study of Life Sciences (Egenis), University of Exeter, Exeter, UK
e-mail: b.smart2@exeter.ac.uk

G. Schiemer (ed.), *The Legacy of the Vienna Circle*, Vienna Circle Institute Yearbook 30, https://doi.org/10.1007/978-3-031-80568-4_9

Keywords Karl Popper · Scientific pluralism · Critical rationalism · Truth · Verisimilitude · Demarcation · Rational reconstruction · Disunity of science · Justificationism · Plurality

9.1 Introduction: Popper's Caricature—Or How Popper Can Be Lumped Together with Monism

> Popper's falsificationist criterion of scientificity has, however, been almost universally rejected. (Dupré 1993, 230).

Since Thomas Kuhn, philosophy of science has had antagonistic attitudes towards Karl Popper and his philosophy (Laudan 1983; Giere 1988; Godfrey-Smith 2003; McGrew et al. 2009). This attitude is particularly prevalent for proponents of scientific pluralism and pluralist-adjacent approaches (Hacking 1983; Dupré 2012, 21–39). Reinforcing this attitude is the thought that Popperian philosophy has become outdated, if not supplanted by 'scientific pluralism', a view that is widely regarded and supported by many contemporary philosophers of science. For instance, Peter Godfrey-Smith suggests that: "the adjective [Popperian] is used for very general ideas that, according to most current philosophers, Popper did not develop convincingly" (2016). Although "[Popper's] detailed account is often seen as attractive on first impression, [it is thought to be] full of holes that become bigger rather than smaller as discussion continues" (ibid. 104).

Since Popperian philosophy is thought, along with many traditional approaches, to have been justifiably relegated to mere historical significance, philosophers of science have found little reason to reappraise it. Nevertheless, this opposition to Popperian philosophy is complicated by the fact that many contemporary philosophers of science are not always explicit or forthcoming about the rationale for their antipathy towards Popperian views. Opponents may assume that decisive arguments have been provided already or else they suggest that present discussions have moved on, rendering irrelevant any doubt that may be cast on the decisiveness of prior argument (e.g. Laudan 1983; contested by Pigliucci 2013 and Lutz 2011).

Some authors, however, have already begun to reevaluate the current epistemological relevance of specific parts of Popper's work, including: his sceptical realism, the role of eliminative inferences, the epistemic value of risk-taking, the possibility of pseudo-evidence via observations in an irrelevant research context, and epistemic processualism (Godfrey-Smith 2016). Sometimes, Popper is even regarded as a predecessor of social epistemology (Longino 2019). We wish to contribute towards this re-evaluation by examining the compatibility of scientific pluralism and Popper's philosophy.

Before we commence our investigation in detail, it is important to note that 'pluralism' is often used as an indicator for a collection of loosely related stances popular in current philosophy of science, particularly driven by the practice turn (Soler et al. 2014). Pluralism comes in many flavours. Scientific pluralism might be metaphysical or epistemological, normative or descriptive, supportive of integration or against it (Veigl 2021). In what follows, we thus refer to 'pluralism' as this loosely connected set of positions, but we pick particular pluralist positions, advanced in the philosophy of science literature, to illustrate how they relate to Popper's work. As we will show, by examining particular pluralist positions, it is possible to highlight the ways in which there are pluralist tendencies in Popper's work, specifically concerning the benefits of a plurality of theories, models or explanations. At the same time, however, we can extract particular forms of pluralism, specifically when it regards scientific methodology, as something that constitutes a point of friction within pluralism itself.[1]

9.2 Pluralists Against Popper

Despite the general aims of this paper, we acknowledge that, in some cases, it is reasonable to think that particular features of Popperian philosophy are monist, not least given some of the explicit monist-sounding remarks of Popper himself. One explicitly monist feature of Popperian philosophy is a commitment to a single principle or criterion for the demarcation of science, which then enables the distinction of science and pseudoscience. Popper is quite clear about his view: a scientific conjecture, hypothesis or claim is only scientific insofar as it is possible to falsify it (*a posteriori*) by a test that has an empirical basis; pseudoscientific claims cannot be falsified by such tests, not even in principle (Popper 1934; cf. Pigliucci 2013). Contrarily, many pluralists challenge the supposition that there is any unity of method to the sciences, which is to say that there is no commonality beyond trivial or superficial features, like the use of elementary mathematics or common language (Suppes 1978).

In another place, Popper argues against a version of pragmatism (or 'instrumentalism') which claims that scientific theories are only instruments for prediction, practical reasoning or practical action (Popper 1967, §8; cf. Popper 1979). Accordingly, we either cannot (or should not) aim at truth in the sense of a pure knowledge of the world, and we are deceived if we think that theories afford explanations or understanding of such (see: Suppes 1967; Love 2014). In this way, scientific theories are compatible with a range of different motivations and aims, though never towards a single depiction of nature. Pragmatic approaches may lead to pluralism, and Popper appears to set himself as its opponent.

We agree that there are features of Popperian philosophy that are fairly characterised as monist, though we may quibble about how these features are presented. However, the two positions just described are distinct epistemological problems and two particular features of Popperian philosophy. Nevertheless, moderate forms of

[1] This is something that has been correctly pointed out by many advocates for pluralism (e.g. Mitchell 2003; Nathan 2019).

pluralism may also fit the same monistic characterisation regarding one or both of these problems. To insist that any thinker is wholly monist or pluralist regarding their views on science is, therefore, to court a paradox.

From the perspective of some scientific pluralists, it may appear as though Popper is closer to the projects he opposed than to contemporary views—the logical positivists, for instance (Popper 1979; cf. Hacohen 2016). During his lifetime, Popper was accused of exemplifying 'positivism' by sociologists and philosophy professors in a muddled dispute involving Theodor Adorno and Jürgen Habermas (Maus and Fürstenberg 1970; Popper 1970a; Strubenhoff 2018). As Ian Hacking (1983) commented, such a comparison is wildly inaccurate. There are important differences between 'positivism', 'logical positivism', and 'Popperianism'. However, since Popper also defends the rationality and unity of science, there are some similarities also.

More importantly, specific differences between Popper and the members of the Vienna Circle do not seem to matter to Popper's opponents (Hacking 1983, 43–44). Indirect antipathy to Popperian philosophy appears to be supported when it can be lumped together with "received" or "traditional" views within philosophy of science. Since pluralists (and pluralist-adjacent approaches) set themselves against these views, framed as an alleged "revolution" or "intellectual turn", then the aforesaid lumping argument provides rhetorical support (see: Giere 1988; D'Amico 1989; Kindi and Arabatzis 2012; Soler et al. 2014; Lidgard and Nyhart 2017). In what follows, we aim at overcoming the rhetoric of the opposition of the two **P**s (Popper and pluralism). First, we will examine two flavours of scientific pluralism in detail. Then, we shall examine points of commonality and friction between these two pluralisms and Popperian philosophy.

9.2.1 Radical Pluralism Against Popper

While there are many flavours of scientific pluralism, we shall rely on a distinction introduced by Kellert et al. (2006): "radical" vs. "modest" pluralism. Kellert and colleagues describe "radical" scientific pluralists as those, whose pluralism invites "relativism." On the one hand, the authors classify "constructivist" positions through this label. On the other hand, they also believe that there are realist versions of the thesis that there is "an indefinite number of theories, the only constraint being human ingenuity" (ibid. xiii). What Kellert and colleagues criticise about such positions is that these pluralists do not provide strict enough criteria for theory choice and thus would be unable to demarcate "bad," or simply "unscientific" theories from the pluralist's repertoire. Another version of "radical" scientific pluralism they mention are versions of pluralism that support a particularly strong metaphysics when it comes to the relation of a plurality of theories with what the world is like, where "any kind term that has a role in a system of understanding refers to a kind" (ibid. xiii). In short, radical scientific pluralism refers to those positions that in one way or the other admit an "anything goes" attitude—either they admit "anything goes" with regards to which scientific theories are acceptable, or they admit "anything goes" with regards to what the world might be like.

In this section, we shall concern ourselves with "radical" positions that are somewhat "relaxed" when it comes to admitting theories to the pluralist's table, while still upholding a realist attitude—these theories maximise our contact with exterior reality. Let us, however, reemphasize here, that we are using the radical-modest distinction as a heuristic tool to structure different pluralist positions; "radical" is a pretty vague term and we do not ourselves commit to the idea that there is something particularly radical about the positions discussed. We believe, however, that particularly this kind of pluralism presents a great case for arguing that many Popperian themes are still present in "millennial" pluralism. We shall focus on Hasok Chang and his pluralist manifesto *Is Water H2O?* as a representative of this form of pluralism.[2]

Quite generally, this branch of radical scientific pluralism can be described by emphasing a plurality of different theories and, specifically, underlining the "incommensurability" of different theories, sets of epistemic practices, viewpoints, explanations, etc. This means that authors supporting "radical but realist" pluralism emphasise that admitting contradictory theories within the scientific enterprise is beneficial for furthering scientific knowledge. Hasok Chang coins his position 'active normative epistemic pluralism'—an "active stance committed to the cultivation of plurality" (2012, 253). Different theories, or, in Chang's vocabulary, 'systems of practice', might be either incommensurable, because their respective answers to a question might not be intelligible by the respective other, or they might be, in fact, contradictory, meaning that they provide different answers (e.g., "yes" and "no") to a particular question asked (Kusch 2016).

Furthermore, radical scientific pluralists usually support a "democratic" ideal of science, meaning, that they metaphorically appeal to the ideal of liberal democracy, where also unorthodox, "radical," or outlier positions can find a voice and can be represented in a way matching their popularity within a population, to argue that scientific discourse should proceed similarly in this respect: While "mainstream" funding bodies do not need to change their way of operating entirely, they should also make space for "minority" scientific endeavours, whose predictions might lie gravely in conflict with theories of certain mainstream ideas:

> Without pretending that the current forms of pluralist liberal democracy are anywhere near perfect, we should also acknowledge that they are protecting us from far worse excesses. There is a simple and crude pluralist lesson: at least have a two-party system, not a one-party system; (Chang 2012, 264).

In this vein, proponents—borrowing again from a different tradition of thought—argue that the "monopoly" of any particular scientific theory is always problematic. Instead, one should have several alternatives around.[3] Pointing towards heuristics,

---

[2] Another example would be Feyerabend's *Against Method* (1975).

[3] Like all such comparisons, this lesson from economics should be taken with a grain of salt. Single vendors in a given context ("monopolies") are not necessarily problematic. A monopoly in this

they argue that we can never be certain that a particular theory (no matter how successful) "got it right." Chang phrases this in individualistic terms—appealing to the virtues of prudence and humility.

Yet, prudence is not the only reason for entertaining a multiplicity of different theories. "Radical but realist" pluralists also argue that alternative theories have the potential to make accessible more empirical facts—every theory has a particular focus on certain things, phenomena, and/or theoretical entities it finds important. Chang identifies this problem in what has been coined '*Kuhn Loss*'—that after a paradigm change, the new paradigm will never be able to account for all the empirical facts that the old paradigm could account for (i.e. a shift in emphasis is happening). Chang suggests a "conservationist" pluralism here:

> retain previously successful theories and paradigms for what they were (and are) still good at, and add new theories and paradigms that will help us make new and fresh contacts with reality. (2012, 224).[4]

And

> What is really wrong in the pessimistic induction scenario is the notion that a successful theory should be rejected if another successful theory comes along and seems to declare the old theory to be false. (2012, 227).

Thus, since one's investigative lens will be limited through the theoretical underpinning one adopts, it is beneficial to admit more than one theory, so more empirical facts can be identified. By these facts, of course, it will in turn be possible to assess different theories. Finally, different, alternative, theories might also interact with each other in fruitful ways. Hasok Chang lists 7 benefits of adopting "radical but realist" pluralism, divided in benefits of "tolerance" ("hedging the bet," "division of domain," "satisfaction of different aims," "multiple satisfaction") and "interaction" ("integration," "co-optation," "competition"). To summarise, radical but realist scientific pluralism is characterised by the idea that a plurality of theories makes accessible new empirical facts, and thereby keeps all approaches "on their toes," which is beneficial for all science. Not every theory, however, needs to be compatible with all empirical facts available.

sense might neither charge monopoly prices nor be able to block market entry for truly superior competitors in the future. A monopoly company might just be able to provide a superior good very cheaply. Are subsidies for inferior companies at the cost of consumers and tax-payers always judicious, just for plurality's sake? An analogous question can be asked with regards to scientific theories that are extremely successful and do not, in principle, block even better theories which might be devised in the future.

[4]This widespread and vague discourse about "contacts with reality" could be interpreted as a regress to pre-logical-empiricist and pre-Popperian times in philosophy of science as it neglects the open problems of justifying sentences by other entities than sentences.

9.2.2 Moderate Pluralism Against Popper

We now turn to moderate forms of scientific pluralism. Angela Potochnick's (2017) 'limited pluralism' is an example of moderate pluralism, as encapsulated in her *Idealization and the Aims of Science*. Other examples should include Sandra Mitchell's 'integrated pluralism' (Mitchell 2003, 2009) and Catherine Elgin's 'indexical perspectivalism' (Elgin 2010, 2017). Moderates consider scientific research to be best characterised by a productive plurality, resulting from the practices of scientific agents. Despite the diversity or heterogeneity of science, there is something, minimally speaking, that unifies science and/or demarcates it from its negation. Additionally, the epistemic progress of science is not strictly captured by increasing truth of scientific representations.

Like their more radical counterparts, moderates do not think scientific representations are simple mirrors of nature. For Elgin (2010), empirically-based approaches to scientific research often lead to heterogeneity within scientific understanding, partly because any empirical measurement is itself conditional, indexical, and perspectival.[5] Potochnik (2017) argues that, in science, idealisation is rampant, largely unchecked, and epistemically beneficial. Causal factors that are important can be advantageously idealised when they are not important to a researcher's immediate interests. Scientific research reflects a variety of epistemic and pragmatic aims, which is best pursued autonomously (Potochnik 2010). Such circumstances lead to a proliferation of idealisation with little done to centrally control, regulate, or limit their influence on scientific representations. Contrastingly, Mitchell (2003, 2009) argues that complex systems phenomena require scientists to collaborate in forming integrated multilevel explanations. The competition between mutually exclusive theories surely describes one aspect of the structure of science, but there is more to it. A variety of scientific explanations for one and the same phenomenon may complement each other, and they may even be compatible with each other.

For moderates, there is a meaningful epistemological boundary between science as a systems phenomenon and its negation, but science is not as uniform or homogenous as naïve versions of monism suggest.

Some moderates promote a plurality of competing theories as a temporary stage in the growth of science: they support competition insofar as it speeds the process towards the acceptance of the better and more unifying theory (Friedman 1974; Kitcher 1981). Moreover, if two (or more) competing theories are uncertain, then moderates may argue that it is rational to entertain both until empirical evidence can contribute to a decisive judgement.

For moderate pluralism, the progress of science is more adequately explained in terms different from the truth of epistemic content—or their increasing verisimilitude ["truth-likeness], which is how scientific progress is characterised for Popper (see e.g. Cartwright et al. 2022). Regarding scientific epistemology, moderate pluralists deny that 'true belief' is basic and the only epistemic good, a position that

[5] This is an idea that Elgin builds on from van Fraassen, albeit in a more moderated form.

Pritchard (2010) refers to as *T-monism* about epistemic value. Similarly, moderates oppose what Richard Levins (1966) referred to as the "naïve" and "brute force" approach to building scientific models—constructs that are a faithful, one-to-one, reflection of the phenomena themselves (cf. Potochnik 2017, 42–50). Such models would require an enormous, and perhaps inconceivable, amount of complex details. Popper appears to fit with monism in this way. He insists on the importance of theories to the logic of scientific discovery, namely in structuring observations.[6] This is further supported by Popper's claim that science grows with respect to a greater accumulation of verisimilitude, along with his own endorsement of Tarski's (1944) correspondence theory of truth. In any case, these are views that pluralists contest.[7]

Moderate pluralism is a two-pronged attack against such monist views about science. On the one hand, the presence of so many idealisations or false representations within science should lead us to reconsider the value of 'truth' to practice, and the basic role it is purported to play. On the other hand, a serious consideration of the pragmatics of scientific understanding appears to lead to the acceptance of there being an irreducible relationship between scientists (as epistemic agents) and the representations they form. Consequently, there is no space for thinking that scientific truth or scientific objectivity constitutes a non-perspectival representation, which functions to "… carve the mind-independent world at its mind-independent joints" (Elgin 2017, 151. Also see: Giere 1994; Chakravartty 2010).

Scientific understanding is often expressed in ways that are patently not true, many of which do not even purport to be true. Elgin (2017) refers to this category of representations as '*felicitous falsehoods*'. Familiar examples in science include: curve smoothing, *ceteris paribus* claims, stylised facts (i.e., approximately true generalisations that admit of exceptions when specificity is increased), idealisations, as well as *a fortiori* arguments that are made from limiting cases.[8] As Elgin argues, to

[6] Note that we jumped swiftly from the modern "true belief" to the critical rationalist and logical empiricist "sentences intersubjectively and tentatively accepted as true, following methodological rules", and from "models" to "theories" – this is due to most scientific pluralists rejecting the syntactic view regarding the relation of theories and models, following Ron Giere's (1999) philosophy, denying that theories and models are strictly distinguishable.

[7] Alan Love (2014), for instance, challenges the assumption that theories are necessary for the practice of scientific research and basic to its organisation. Love cites Kuhn (1962/2012), along with Oppenheim and Putnam (1958), as his examples, but the same arguments can be sent Popper's way. Traditionally, philosophers have assumed that each programme of scientific research is structured, organised, or guided by a leading theory (e.g. a paradigmatic explanation, an axiom system, or a conceptual framework, etc.). Love argues that some scientific fields (e.g. developmental biology) are structured erotetically, and thereby on a different basis than T-monists claim. Having said that, any (perhaps tentative) answer to a question is a theory.

[8] Similarly, for Potochnik, there are many scientific practices that do not clearly appear to resemble the successive approximation of truth or the increasingly accurate representation of phenomena, with the purpose of explaining or predicting those phenomena. In her view, scientific practices involve a rampant and unregulated use of idealisation (Potochnik 2017, 47–48). In each instance, *productively idealising* brings about an epistemic advantage from misrepresenting a causal pattern, which is of scientific interest. There are many reasons to idealise, with some reasons occurring in combination. Some involve the justification of a temporary idealisation, whereas others involve

the extent that an apparently successful scientific theory makes use of felicitous falsehoods, the T-monist is compelled either to deny its success or to explain how a strictly untrue theory is instrumental towards bringing about a true (or truer) theory. The second approach works like a bet: "...that science will eventually discover the resources to eliminate [the felicitous falsehoods of a given theory] without replacing them with other felicitous falsehoods" (Elgin 2017, 31f.). The difficulty with the T-monist view is that, although scientists expect current models and idealisations to be improved as science progresses, they do not expect science to outgrow the use of such devices. Therefore, the epistemic value of truth is far too narrow to explain the role and function of felicitous falsehoods in science.

9.3 Millennial Popperianism: Popper as Partly Pluralist

We believe that Popper and his philosophy have often been unfairly misrepresented by contemporary philosophers of science. In particular, we shall show two characteristics of Popper's view that are more pluralist, albeit a moderate one, than his reception often thinks. First, plurality is important for the pragmatics of (rational) science. Second, truth is not the ultimate (epistemic) aim of scientific activity, since understanding our scientific problems (and the phenomena they concern) is much more important.

9.3.1 Pro Plurality

Radical pluralists are staunch proponents of a plurality in science. For most pluralists, this stance extends to many different types of scientifically relevant entities such as theories, explanations, practices, research programmes, solutions to problems, models_1 (as in representing a target system), models_2 (as in non-linguistic objects for which theories are true or false), perspectives, standpoints, scientific languages, and methodologies.[9] At least for the first seven of these types of entities, radical pluralists are arguably even surpassed by Popper in his unwavering plea for plurality.

Over and above a normative approval of plurality, Popper's view of the history of science descriptively differs from Kuhn's insofar as Popper emphasises that "there

justifying an idealisation permanently. These reasons reflect epistemic limitations that arise as features of either our complex world or the epistemic agency of scientists and/or their audiences (cf. Potochnik 2016).

[9] See e.g. Wylie (2015). Systematic developments of standpoint pluralism about the sciences yet have to be developed. Dealing with a plurality of standpoints and at the same time capturing intersectionality, as well as what to do about androcentric scientific communities poses a challenge (see Veigl forthcoming).

was, ever since antiquity, constant and fruitful discussion between competing dominant theories" instead of one dominating paradigm (Popper 1970b, 55). Put more confrontationally, "Kuhn is mistaken when he suggests that what he calls 'normal' science is normal" (ibid., 53).

Popper's encapsulates his embrace of plurality when he presents what he takes to be the three requirements for the success of science and other endeavours: "...success depends mainly on three conditions, namely, that sufficiently numerous (and ingenious) theories should be offered, that the theories offered should be sufficiently varied, and that sufficiently severe tests should be made" (Popper 1969, 421).[10]

Having a large pool of quite different theories at disposal increases the chance that a particularly useful one and a particularly truth-like one will be identified. Just like radical pluralism, Popper's approach aims to maximise the confrontation of theories with "reality". Critical tests are pivotal for critical rationalists, first and foremost in the form of attempts to falsify, but also inviting other forms of criticism. Admittedly, Popper might emphasise the possibility to (provisionally) weed out some theories, which is less radical than "anything goes pluralists", but even radicals usually want to limit admitting any nonsensical theory imaginable. In fact, most if not all "radical" scientific pluralists spill considerable ink on how to "keep the foul fruits out."

A rare consensus between Popper and Laudan (1981) can be identified here, namely that the history of science teaches us that some highly successful scientific theories did not resist criticism, and that this should instil humility with regards to the fate of current theories, which might seem irrefutable momentarily—an epistemic virtue that Chang also emphasises. Therefore, it is prudent to not (monistically) bet all the money on just one horse:

> There will be contingencies, so we need to have a kind of science that has some capacity for handling surprises, so that if one thing fails not everything else fails as a consequence. (Chang 2012, 257).

Considering the anti-justificationism of critical rationalist epistemology, plurality is perhaps even more of an imperative for Popperians than for many millennial pluralists. According to critical rationalism, a theory which has withstood many severe tests (so far) is no more secure from falsification and rejection than anyone's novel conjecture of the moment. For a theory, past successes do not enhance certainty. Confirmation as an increase in absolute or partial certainty is rejected conceptually and pragmatically. The aim of Popperian testing is solely to provisionally classify a theory as wrong in case it fails the test. Even if the theory passes many tests, i.e. is well corroborated, it remains purely conjectural knowledge. Consequently, Popperians need to give many new theories a chance even if current dominant theories have an excellent track record.

---

[10] That the theory 'better be ingenious' is added in the reformulated 1963 version of the paper (included in Popper 1969). Otherwise, the three conditions had already been stated in the first version of the paper (Popper 1940, 404).

By contrast, since most millennial pluralists adhere to some form of justificationism, they face a constraint to their pluralism. A well-confirmed theory is more likely to be successful in the future (and to be true); therefore, it arguably deserves attention and funding, pursuant to its high degree of confirmation. Unlike Popper's critical rationalism, justificationist mainstream methodology will happily select favourite theories based on expected future success. However, pluralism requires that underdog theories are attentively considered which is tantamount to overrating their prospects.[11] Perhaps a Popperian alliance between pluralism and anti-justificationism would ultimately be a more coherent combination.

In another respect, Popperian pluralism is more superficial than some versions of millennial pluralism though. While some authors conceive of a plurality of theories and approaches as mirroring a fundamental scattered variety in the real world (Dupré 1993; Cartwright 1999), Popper is agnostic with respect to the regularity of the universe and the disorder of things. He merely urges scientists to search for laws and order. In this sense, Popperian pluralism has no ontological motivation but is based on pragmatic and epistemological considerations only. Thus, Popper is more aligned with Chang, rather than "metaphysical" pluralists such as Dupré and Cartwright.

A critical rationalist is more likely than a millennial pluralist to mention that, in a given language L, there is only one complete theory which is true (in a given model that stands as proxy for the "real world"). For every sentence φ of L, this theory contains either φ or not-φ. Of course, Popper would hasten to add that we can never know whether we found that theory, not even whether we are more or less likely to have found it. Thus, we should keep testing, comparing the explanatory power of alternative theories, and comparing their capability to solve theoretical and practical problems. This pluralist endeavour is fueled by pragmatic and anti-justificationist considerations, not by strong ontological commitments.

Finally, as hinted at earlier, Popper's pluralism extends to some types of entities relevant to science but not to others. Most importantly, ever since Neurath, many pluralists criticise Popper for his monism about scientific methodology (Neurath 1935/1983; Neurath and Carnap 2019; see also Kusch this volume, Linsbichler 2021, 2022). For Popper, no viable alternatives to falsificationism have been found.[12]

[11] Elgin (2017) is well aware of this pitfall and aims to avoid it.

[12] The term "falsificationism", which Popper did not use, is misleading as it wrongly insinuates that empirical testing is the only way to evaluate theories. Popper endorses comparative appraisals of theories as to whether they are able to solve given problems. The alternative term "fallibilism" which some critical rationalists prefer is problematic for different reasons: Most philosophers use "fallibilism" without the idea of purely conjectural knowledge.

9.3.2 Truth and Other Objectives

Pluralists often portray Popper as a T-monist and criticise this position. They maintain that Popper values truth as the only objective of science and that he consequently presents a distorted picture of scientific practice and scientific progress. While Popper's methodological considerations do emphasise the search for truth as a pivotal regulative idea in science and beyond, regulating among a vast variety the choice of methods implemented in order to cope with the problem of evaluating the problem-solving capacity of (scientific) theories and ideas, we want to qualify and mitigate such characterizations of Popper as a T-monist in several respects.

First, most of Popper's writings address *epistemological*—as opposed to psychological, sociological, or historical—problems, and thus often treat theories from the narrow perspective as if they were solely attempts to separate true from false sentences (Popper 1934/1959, ch1). This by no means excludes studies of theories and their usage from other perspectives in which truth does not play a privileged role.

Second, even if there is a focus on truth, Popper hastens to clarify that the objective of science is not to find any true sentences, but *interesting* true sentences. Interesting truths are, according to Popper, sentences claiming to be true and having a high informational content and explanatory power but a low probability in terms of probability calculus (Popper 1960/1, p. 311, see also sect. XIV). Note that the probability space chosen and the phenomena worth explaining are relative to the value judgments of the scientist and her community.

Third, although Popper harshly condemns instrumentalism, he acknowledges that scientific theories can serve many purposes.

> *...of course, that a scientific theory may be applied to all sorts of practical problems, either at once, when first invented, or at a later date. And we therefore have no objection to the assertion that all scientific theories are instruments—either actual or potential instruments. But we assert that they are not merely instruments. For we assert that we may learn from science something about the structure of our world: that scientific theories can offer genuinely satisfying explanations that can be understood and so add to our understanding of the world.* (Popper 1967).

The final part of the quote hints at the intimate relationship that Popper, not unlike Potochnik, postulates between truth and understanding (also see: Popper 1949, p. 173; 1963, p. 98–99; 1972, p. 35). Truth and understanding are not the only objectives of science that Popper recognizes. Especially once he had abandoned the idea of a quantifiable notion of 'degree of verisimilitude', keeping the notion as a mere heuristic, new quality criteria for the evaluation of scientific theories become more prominent. In comparison to their alternatives, good theories make more precise assertions, take account of more facts, describe facts in more detail, have passed more tests, suggest new innovative tests, and unifies hitherto unconnected problems (Popper 1960/1). Admittedly, these other objectives of science are, in Popper's account, closely connected with verisimilitude and thus also with truth. Since this link cuts both ways, a converse point can and has been raised against Elgin and other millennial pluralists.

Fourth, Warenski (2019), Nawar (2019), Lawler (2019), and indirectly also Rice (2019) and Le Bihan (2019) object to Elgin's prioritisation of understanding as the main objective of science.[13] These critics argue that understanding, even if it is non-factive as in Elgin's case, crucially relies on (or can be elucidated by) the notion of truth.

In summary, millennial pluralists tend to overstate the role of truth in Popper and, arguably, underestimate the epistemic role of truth in their own approaches.

There is one more bone to pick in this section: Popper's arguably one-sided appropriation of Tarski's notion of truth. Popper reveres Tarski as rehabilitator and proponent of a correspondence theory of truth, with sentences being true if they correspond to facts in "the real world". Yet, it is at the very least questionable whether Tarski's semantic notion of truth is actually committed to such a correspondence theory. The list of counterarguments comprise, firstly, of the fact that Tarski (1944) explicitly and correctly denies that his theory either presupposes or implies realism (about "the real world") and, secondly – and so far mostly disregarded in pertinent discussions – of the relativity of truth to a "model" (in modern nomenclature). In the mature version of Tarski's account, a sentence φ of language L is not true or false simpliciter, but true or false in a model M (more specifically, in an L-model M). The bone whether Tarksi is a correspondence theorist in the Popperian sense of correspondence to "the real world" or whether Popper presents a strategically contorted reconstruction of Tarski's position has been gnawed at a lot in the secondary literature.[14] Although some of us hold a clear exegetical position, we need not resolve the issue here.

Suffice to say that adopting a more modest version of Tarskian truth, namely one without realist commitments (concerning "the real world") would be conducive to millennial pluralism. When sentences can be true in a model M_1 and at the same time false in M_2, and perhaps a third (data) model M_3 serving as proxy for "the real world", it becomes possible to accommodate felicitous falsehoods, idealizations, and models while retaining a consistent notion of truth for sentences. The latter is, among other things, necessary for evaluating the soundness and validity of arguments (as standardly defined). If scientists and philosophers want to make arguments and draw inferences, sentences and at least such a modest notion of truth are all but indispensable (Linsbichler 2023). To be sure, this proposal is a far cry from the absolute notion of truth as correspondence with "the real world" desired and evoked by Popper.

---

[13] Although Popper does not systematically develop his concept of 'understanding', it is obvious that Popper and Elgin certainly employ quite different notions of 'understanding'.

[14] For the initial charge, see Haack (1976). For a brief reaction, see Popper (1976).

9.3.3 Chang Acknowledges Partial Popperian Pluralism

There is one outstanding exception to millennial pluralists swiftly dismissing critical rationalism as antagonistic to their own position. On several occasions, Chang carefully fathoms out the relationship of his own pluralism to Popper's philosophy of science. Chang discusses possible differences as well as commonalities, very much in line with our approach in this chapter. On the one hand, Chang partly sees falsifiability as a way of "maximising contact with reality," a commitment that is one characteristic of his pluralism. On the other hand, it is with Popper's notion of how science progresses that Chang has particular issue: he thinks it would lead to "theory monoculture."

'Active realism' to Chang means the commitment towards maximising contact with reality, meaning to admit several, "systems of practice" that might be in conflict with each other, in order to make accessible a maximum number of empirical facts. Chang identifies commonalities between his approach and Popper: "Karl Popper's injunction for scientists to seek higher falsifiability and more severe tests can also be seen as a demand for more contact with reality" (218f.) In a footnote, he continues: "If we can turn the focus away from true/false hypothesis-testing to a broader view of various epistemic activities that can serve as tests of theories, there is actually not so much difference between what Popper recommends and what I advocate" (ibid.).

Learning from reality also requires situations in which our expectations are likely to be contradicted, which Chang identifies as the "basic intuition underlying Popper's requirement of severe tests." (221). Chang also emphasises how Popper in his debate with Kuhn emphasised that a plurality of theories is required during periods of normal science, where one theory could even become strong enough to effect a paradigm change (258), arguing that "too many dollars may chase too few ideas" (1981, 96).

Another connection Chang identifies, is the rejection of dogmatism: "I can heartily endorse the invectives from Popper and Lakatos against "pseudoscience" with its striving to defend its cherished beliefs at all costs, discarding observations that go against those beliefs and shrinking from making experiments that have a chance of refuting them" (2012, 222). However, Chang identifies a departure point in that both "pseudoscience" and accepted scientific theory may become dogmatic, if they are not challenged by other, contradictory, or incommensurable theories. He identifies the problem here with Popper's acceptance of a "standard" version of scientific realism, which lies in the correspondence notion of truth, and "which assumes that there is only one truth about any given bit of reality" (223). Hence, Popper's critical rationalism (may) ultimately lead to theory monoculture–the reign of the theory with maximum contact to reality. It is exactly here, where Chang believes that standard scientific realism will slide into dogmatism as well.

Popper was being an active realist when he demanded that each new theory should have higher empirical content than its predecessor, but he added an unreasonable demand for continuity which was quite monist in its flavour: "a new theory, however revolutionary, must always be able to explain fully the success of its predecessor" (1975, 94). Popper's demand may have been met in a few evocative moments in the history of physics, such as the transition from Newtonian mechanics to special relativity, in which there was a complete inclusion of the old phenomena in the new paradigm in the numerical sense despite semantic incommensurability. But how many instances of scientific development go like that?" (223). In a later footnote, Chang concedes however: "In a more detailed statement on what it means for one theory to "correspond better to the facts" than another, [Popper 1969] seems to avoid such a strong continuity requirement…" (Chang 2012, 232).

As discussed above, Chang supports "conservationist pluralism"—to retain those programs that were successful in the past and add new ones to make new contacts with reality (224). Thus, Chang advocates a different picture of how knowledge is to grow:

> The cumulative growth of scientific knowledge, beyond what is straightforwardly possible within each system, is a consequence of a mostly unspoken conservationist pluralism by which we keep various systems of knowledge alive. What we have is an accumulation of system- dependent knowledge, which can only happen through an accumulation of systems. (2012, 218).

and

> Popper in his youth seems to have been seduced by a false asymmetry between the certainty of falsification and uncertainty of verification. As long as Popper continued to claim that he had solved the problem of induction, he tied himself to the unproductive search for certainty. (The Popper who used falsifiability as a demarcation criterion did not have this problem.) Popper denounced perfectly legitimate inductive methods for their lack of certainty, while in the end admitting that falsification lacked certainty, too, and bringing in "corroboration" as an uncomfortable proxy for inductive confirmation. (2012, 244).

Contrary to Chang's interpretation, however, one could argue that the rejection of certainty and the provisionality of falsification are key components of Popper's early writings as well. At some point between 1930 and 1933, he already stated: "Science, however, requires not certainty but only successful predictions" (Popper 1979, 744; see §46, i.e. 598–631, for more extensive discussion). His *Logik der Forschung,* published in 1934, ends with a passionate declamation for the uncertainty and revisability of all scientific knowledge:

> The old scientific ideal of epistēmē—of absolutely certain, demonstrable knowledge—has proved to be an idol. The demand for scientific objectivity makes it inevitable that every scientific statement must remain tentative for ever. It may indeed be corroborated, but every corroboration is relative to other statements which, again, are tentative. Only in our subjective experiences of conviction, in our subjective faith, can we be 'absolutely certain'. [footnote deleted] With the idol of certainty (including that of degrees of imperfect certainty or probability) there falls one of the defences of obscurantism which bar the way of scientific advance. For the worship of this idol hampers not only the boldness of our questions, but also the rigour and the integrity of our tests. The wrong view of science betrays itself in the craving to be right; for it is not his possession of knowledge, of irrefutable truth, that makes the man of science, but his persistent and recklessly critical quest for truth. (Popper 1934/1959, 280–281).

In any case, Chang's analysis of Popper (which is, at times, hidden in footnotes) agrees with many of our observations regarding Popper's position on a plurality of theories. Pleas for the necessity of plurality share many correspondences between Popper and scientific pluralists such as Chang. Where Chang locates differences is with regards to his outlook on scientific pluralism. Yet, as we have argued in Sect. 9.3.1, Popperian anti-justificationism might be a fruitful, yet so far unexamined, avenue for pluralists concerned with scientific progress.

9.4 A Brief Genealogical Outlook

In this article, we have tried to re-examine connections between Popper's philosophy and recently formulated scientific pluralist positions. We have argued that so far, most scientific pluralists have overstated the differences between their philosophies and Popper's theorising. We, however, also do not want to make the opposite mistake and overemphasise similarities. Consequently, we want to acknowledge that several differences remain, such as views about scientific progress. In the future, it might be fruitful to employ a "stance" perspective, as has become prevalent in philosophy of science when examining a set of complicated views, to further extrapolate commonalities and differences. In addition, despite showing several underappreciated commonalities, we have not delved into the genealogy of these commonalities, and thus not considered the question whether Popper's philosophy has been a direct influence on the relative pluralisms. To address such questions, Paul Feyerabend's pluralism might provide a fruitful starting point.

Acknowledgments A previous version of this talk was presented at the *120 Years since the Birth of Karl Popper Conference* at the University of Sofia, Bulgaria. We thank the participants, as well as Karl Milford, for helpful discussions. The authors are grateful to Georg Schiemer and Friedrich Stadler for their circumspect editorial work.

Funding Alexander Linsbichler's research is funded in whole, or in part, by the Austrian Science Fund (FWF) (grant number ESP 206-G; grant-DOI 10.55776/ESP206).

References

Cartwright, Nancy. 1999. *The Dappled World. A Study of the Boundaries of Science*. Cambridge University Press.

Cartwright, Nancy, Jeremy Hardie, Eleonora Montuschi, Matthew Soleiman, and Ann C. Thresher. 2022. *The Tangle of Science: Reliability Beyond Method, Rigour, and Objectivity*. Oxford University Press.

Chakravartty, Anjan. 2010. Informational Versus Functional Theories of Scientific Representation. *Synthese* 172 (2): 197–213.

Chang, H. 1981. The Rationality of Scientific Revolutions. In *Scientific Revolutions*, ed. Ian Hacking, 80–106. Oxford University Press.

———. 2012. *Is Water H2O?: Evidence, Realism and Pluralism*. Springer Science and Business Media.
D'Amico, Roberto. 1989. *Historicism and Knowledge*. Routledge.
Dupré, John. 1993. *The Disorder of Things: Metaphysical Foundations of the Disunity of Science*. Harvard University Press.
———. 2012. *Processes of Life: Essays in the Philosophy of Biology*. Oxford University Press.
Elgin, Catherine Z. 2010. Keeping Things in Perspective. *Philosophical Studies* 150: 439–447.
———. 2017. *True Enough*. MIT Press.
Feyerabend, Paul K. 1975. *Against Method: Outline of an Anarchistic Theory of Knowledge*. New Left Books.
Friedman, Michael. 1974. Explanation and scientific understanding. *Journal of Philosophy* 71: 5–19.
Giere, Ronald N. 1988. *Explaining Science: A Cognitive Approach*. University of Chicago Press.
———. 1994. The Cognitive Structure of Scientific Theories. *Philosophy of Science* 61 (2): 276–296.
———. 1999. *Science Without Laws*. University of Chicago Press.
Godfrey-Smith, Peter. 2003. *Theory and Reality: An Introduction to the Philosophy of Science*. University of Chicago Press.
———. 2016. Popper's Philosophy of Science: Looking Ahead. In *The Cambridge Companion to Popper*, ed. Jeremy Shearmur and George G. Stokes, 104–124. Cambridge University Press.
Haack, Susan. 1976. Is it True What They Say About Tarski? *Philosophy* 51 (197): 323–336.
Hacking, Ian. 1983. *Representing and Intervening: Introductory Topics in the Philosophy of Natural Science*. Cambridge University Press.
Hacohen, Malachi H. 2016. The Young Popper, 1902–1937: History, Politics and Philosophy in Interwar Vienna. In *The Cambridge Companion to Popper*, ed. Jeremy Shearmur and George G. Stokes, 30–68. Cambridge University Press.
Kellert, Stephen H., Helen E. Longino, and C. Kenneth Waters. 2006. *Scientific Pluralism*. University of Minnesota Press.
Kindi, Vasso, and Theodore Arabatzis. 2012. *Kuhn's The Structure of Scientific Revolutions Revisited*. Routledge.
Kitcher, Philip. 1981. Explanatory Unification. *Philosophy of Science* 48: 507–531.
Kuhn, Thomas S. [1962] 2012. *The Structure of Scientific Revolutions*. 4th ed. University of Chicago Press.
Kusch, Martin. 2016. Relativism in Feyerabend's Later Writings. *Studies in History and Philosophy of Science Part A* 57: 106–113.
———. this volume. Relativism and Sociology of Knowledge in Logical Empiricism: The Cases of Frank and Neurath. Chapter 8. In *The Legacy of the Vienna Circle*, ed. Georg Schiemer. Cham: Springer Nature.
Laudan, Larry. 1981. A Confutation of Convergent Realism. *Philosophy of Science* 48 (1): 19–49.
———. 1983. The Demise of the Demarcation Problem. In *Physics, Philosophy and Psychoanalysis*, ed. Robert S. Cohen and Larry Laudan, 111–127. Reidel.
Lawler, Insa. 2019. Scientific Understanding and Felicitous Legitimate Falsehoods. *Synthese*. https://doi.org/10.1007/s11229-019-02495-0.
Le Bihan, Soazig. 2019. Partial Truth Versus Felicitous Falsehoods. *Synthese* 198: 5415–5436.
Levins, R. 1966. The Strategy of Model Building in Population Biology. *American Scientist* 54 (4): 421–431.
Lidgard, Scott, and Lynn K. Nyhart. 2017. *Biological individuality: Integrating Philosophical, and Historical Perspectives*. University of Chicago Press.
Linsbichler, Alexander. 2021. Rationalities and Their Limits: Reconstructing Neurath's and Mises's Prerequisites in the Early Socialist Calculation Debates. *Research in the History of Economic Thought and Methodology (RHETM)* 39B: 95–128.
———. 2022. Als Schwurbler Logik und Empirie vertrieben. *Wirtschaft und Gesellschaft* 48 (1): 109–117.

———. 2023. An Ultra-refined Grammar for Interactions: Thoughts on Robert Aumann's Philosophy of Game Theory. *Revue Économique* 74 (4): 635–650.
Longino, Helen. 2019. *The Social Dimensions of Scientific Knowledge*. Stanford Encyclopedia of Philosophy. https://plato.stanford.edu/archives/sum2019/entries/scientific-knowledge-social/. Accessed 1st May 2024.
Love, Alan C. 2014. The Erotetic Organization of Developmental Biology. In *Towards a Theory of Development*, ed. Alessandro Minelli and Thomas Pradeu, 33–55. Oxford University Press.
Lutz, Sebastian. 2011. On an Allegedly Essential Feature of Criteria for the Demarcation of Science. *The Reasoner* 5 (8): 125–126.
Maus, Heinz, and Friedrich Fürstenberg. 1970. *Der Positivismusstreit in der deutschen Soziologie*. Neuwied: Luchterhand.
McGrew, Timothy, Marc Alspector-Kelly, and Fritz Allhoff. 2009. *Philosophy of Science: An Historical Anthology*. Wiley.
Mitchell, Sandra D. 2003. *Biological Complexity and Integrative Pluralism*. Cambridge University Press.
———. 2009. *Unsimple Truths: Science, Complexity, and Policy*. University of Chicago Press.
Nathan, Marco J. 2019. Pluralism is the Answer! What is the Question? *Philosophy, Theory, and Practice in Biology* 11 (15). https://doi.org/10.3998/ptpbio.16039257.0011.015.
Nawar, Tamer. 2019. Veritism Refuted? Understanding, Idealization, and the Facts. *Synthese* 198: 4295–4313.
Neurath, Otto. 1983 [1935]. Pseudorationalism of Falsification. In *Otto Neurath Philosophical Papers 1913-1946*, ed. R. Cohen and M. Neurath, 121–131. Dordrecht: Reidel.
Neurath, Otto, and Rudolf Carnap. 2019. Appendix 2: The 1940–1945 Neurath-Carnap Correspondence. In *Neurath reconsidered: New Sources and Perspectives*, ed. Jordi Cat and Adam Tamas Tuboly, 512–685. Springer.
Oppenheim, Paul, and Hilary Putnam. 1958. The Unity of Science as a Working Hypothesis. In *Minnesota Studies in the Philosophy of Science*, ed. Herbert Feigl et al., vol. 2, 3–36. Minneapolis: Minnesota University Press.
Pigliucci, Massimo. 2013. The Demarcation Problem. A (Belated) Response to Laudan. In *Philosophy of Pseudoscience: Reconsidering the Demarcation Problem*, ed. Massimo Pigliucci and Maarten Boudry, 9–28. University of Chicago Press.
Popper, Karl R. 1934. *Logik der Forschung: Zur Erkenntnistheorie der modernen Naturwissenschaft*. Vienna: Verlag von Julius Springer. English edition: Karl R. Popper, [1959] 2002. *The Logic of Scientific Discovery* (trans. by Karl R. Popper, Julius Freed & Lan Freed). Oxon: Routledge.
———. 1940. What is Dialectic? *Mind* 49 (196): 403–426.
———. 1949. Towards a Rational Theory of Tradition. In *The Rationalist Annual for the Year 1949*, ed. Frederick Watts, 36–55. London: Watts & Company. Reprinted in *Conjectures and Refutations: The Growth of Scientific Knowledge*, 161–182. London: Routledge.
———. 1960/1 (unpublished). Truth, Rationality and the Growth of Scientific Knowledge. Reprinted in *Conjectures and Refutations*. 291–338.
———. 1963. Science: Problems, Aims, Responsibilities. *Federation Proceedings* 22: 961–972. Reprinted in *The Myth of the Framework*. ed. M.A. Notturno, 82–111. London: Routledge.
———. 1967. Models, Instruments, and Truth: The Status of the Rationality Principle in the Social Sciences. Reprinted in *The Myth of the Framework*. ed. M.A. Notturno, 154–184. London: Routledge.
———. [1969] 2002. *Conjectures and Refutations: The Growth of Scientific Knowledge*. 3rd ed. London: Routledge.
———. 1970a. Reason or Revolution? *European Journal of Sociology* 11 (2): 252–262.
———. 1970b. Normal Science and its Dangers. In *Criticism and the Growth of Knowledge*, ed. Imre Lakatos and Alan Musgrave, 51–58. Cambridge University Press.
———. 1972. The Myth of the Framework. Reprinted in *The Myth of the Framework*. ed. M.A. Notturno, 33–64. London: Routledge.

———. 1975. "The Rationality of Scientific Revolutions" (Herbert Spencer Lecture). In *Problems of Scientific Revolution*, ed. Rom Harré, 72–101. Oxford: Oxford University Press.
———. 1976. Is it True What She Says About Tarski? *Philosophy* 54 (207): 98.
———. 1979. *Die beiden Grundprobleme der Erkenntnistheorie*. Tübingen. English edition: Popper, K.R. 2009. *The Two Fundamental Problems of the Theory of Knowledge* (trans. by Andreas Pickel & John Kinory). ed. Troels Eggers Hansen. Oxon: Routledge.
Potochnik, Angela. 2010. Explanatory Independence and Epistemic Interdependence: A Case Study of the Optimality Approach. *The British Journal for the Philosophy of Science* 61: 213–233.
———. 2016. Scientific Explanation: Putting Communication First. *Philosophy of Science* 83: 721–732.
———. 2017. *Idealization and the Aims of Science*. University of Chicago Press.
Pritchard, Duncan. 2010. Knowledge and Understanding. In *The Nature and Value of Knowledge: Three Investigations*, ed. Duncan Pritchard, Alan Millar, and Adrian Haddock, 1–88. Oxford University Press.
Rice, Collin. 2019. Understanding Realism. *Synthese* 198: 4097–4121.
Soler, Léna, Sjoerd Zwart, Michael Lynch, and Vincent Israel-Jost. 2014. *Science After the Practice Turn in the Philosophy, History, and Social Studies of Science*. Routledge.
Strubenhoff, Marius. 2018. The Positivism Dispute in German Sociology, 1954–1970. *History of European Ideas* 44 (2): 260–276.
Suppes, Patrick. 1967. What is a Scientific Theory? In *Philosophy of Science Today*, ed. Sidney Morgenbesser, 55–67. Basic Books.
———. 1978. The Plurality of Science. In *PSA 1978: Proceedings of the 1978 Biennial Meeting of the Philosophy of Science Association*, ed. P. Asquith and Ian Hacking, vol. 2, 3–16. Philosophy of Science Association.
Tarski, Alfred. 1944. The Semantic Conception of Truth and the Foundations of Semantics. *Philosophy and Phenomenological Research* 4 (3): 341–376.
Veigl, Sophie Juliane. 2021. Notes on a Complicated Relationship: Scientific Pluralism, Epistemic Relativism, and Stances. *Synthese* 199 (1): 3485–3503.
———. forthcoming. Research Heuristics, Plurality, and Epistemic Oppression. In: *The philosophy of Helen Longino*, ed. Richardson et al.
Warenski, Lisa. 2019. Epistemic Norms: Truth Conducive Enough. *Synthese* 198: 2721–2741.
Wylie, Alison. 2015. A Plurality of Pluralisms: Collaborative Practice in Archeology. In *Objectivity in Science*, ed. F. Padovani et al., 189–210. Springer.

Part II
General Part

Chapter 10
Otto Neurath and Ludwig Wittgenstein: Money and Philosophy in Vienna

Jordi Cat

Abstract From Marxist analyses to more recent appeals to economic models to address philosophical questions about science in society, relations of money to philosophy prove to be contextual and diverse, partly because of the variety of relations that can be identified, documented, and considered noteworthy, and partly because of the variety of aspects of money and philosophy one can select—material, intellectual, behavioral, symbolic, etc. My comparative argument identifies cross-disciplinary, intellectual, social, and personal connections to the philosophical contributions of two Viennese figures: Otto Neurath and Ludwig Wittgenstein. While I draw attention to different philosophies of money and different roles for their respective attention to money, I also note that nevertheless both shared acquaintance with practical money matters and with theoretical debates and their participants in economics—including their respective fathers.

Keywords Otto Neurath · Wilhelm Neurath · Ludwig Wittgenstein · Karl Wittgenstein · Money and philosophy · Philosophy of economics · Vienna circle · Logical empiricism · Philosophy of language · Philosophy of mathematics · Wittgenstein's philosophy of language · Wittgenstein's philosophy of mathematics

---

Another version of this paper was delivered on 9 November 2023 as the 2023 Vienna Circle Lecture. Both are extracted, with additions, from a longer version in Cat 2024.

J. Cat (✉)
Indiana University, Bloomington, IN, USA

G. Schiemer (ed.), *The Legacy of the Vienna Circle*, Vienna Circle Institute Yearbook 30, https://doi.org/10.1007/978-3-031-80568-4_10

10.1 Introduction

The relations of money to philosophy are contextual and diverse, partly because of the variety of relations that can be identified, documented, and considered noteworthy, and partly because of the variety of aspects of money and philosophy one can select—material, intellectual, behavioral, symbolic, etc. My argument identifies cross-disciplinary, intellectual, social, and personal connections to the philosophical contributions of two Viennese figures: Otto Neurath and Ludwig Wittgenstein.

Wittgenstein's attention to money in his philosophical work appears in his later, post-*Tractatus* period. His later perspective on philosophy and on money is social, but, I argue, the focus is individualistic. Wittgenstein's examples play a demonstrative role in philosophical argument, as illustration and evidence; and this through his attention to individual linguistic and monetary behaviors as social practices. By contrast, Neurath's interest in money and its indirect relevance to philosophical considerations are informed by social goals and considerations of economic theory, methodology, disciplinary identity and intervention. Unlike Wittgenstein's, his contribution to philosophy of science was informed in multiple ways by different interest in and interpretation of money in theory and practice. Both nevertheless shared acquaintance with practical money matters and with theoretical debates and their participants in economics.

10.2 Neurath's Intellectual Development and the Question of Funding

The most socially engaged, vocal, and visible member of the Vienna Circle was Otto Neurath. His father, Wilhelm, was an economist and social reformer, critic of capitalism—not of markets and private property—and proponent of pancartelism. Beyond his economics specialty, his son Otto was exposed to his ethical idealism and commitment to social welfare, in addition to his encyclopedic knowledge and library.[1]

Otto's polymathic education at home and at the university of Vienna extended to the humanities and sciences and subsequently, at the University of Berlin, focused on political economy. He was actively involved in different programs of social reform. His practical and theoretical engagement of political economy became the sources of many of his methodological contributions to the debates within logical empiricism.

Neurath's intellectual activities and contributions may be related to money matters in at least two ways. One is his theoretical, methodological and practical engagement of the economy. The other is the local environment of available resources funding directly or indirectly intellectual life. I begin briefly with the latter.

[1] Johnston 1972, 193.

In 1928 the scientists and philosophers associated with a discussion circle organized by Moritz Schlick formalized their public intellectual promotion of the scientific conception of the world by constituting the Ernst Mach Association. Outside the Association's discussions, members such as Neurath, Frank, and Schlick actively contributed to other local intellectual events and associations.[2] They engaged also in publication projects: a journal *Erkenntnis* (Carnap and Reichenbach, 1930) and two book series, *Monographs on the Scientific World Conception* (Schlick and Frank, 1928–37) and *Unified Science* (Neurath, 1933–39), followed by the *International Encyclopedia of Unified Science* (Neurath, Carnap, and Charles Morris, 1938–70).

All of this activity required a considerable amount of effort and resources. But who funded such activities? As the Austro-Hungarian Empire unraveled, and with it the social order it relied on, a disappearing elite of educated wealthy patrons played a comparatively minor role. Among them was a new minority of socially mobile assimilated Jewish families such as the Gomperzs and Wittgensteins.[3] With fortunes originating in banking and industry, their cultural patronage was distinctively characterized by the value of education, especially in the arts, humanities, sciences, and engineering.

In addition to minor, targeted sponsorship by private groups and individuals, the dominant funding sources were institutional, especially the national and local governments. In the Austrian government, the Ministry of Education paid for university salaries for professors such as Schlick (philosophy) and Hahn (mathematics).

The city's social and then socialist culture prompted the formation of numerous groups promoting exact scientific thinking, adult education, workers education, and school reform. During the so-called Red Vienna period (1919–34), the local government issued a limited but influential number of socialist policies focused on social reform through public housing and education funded by heavier taxes.[4] After several years teaching political economy at the New School of Commerce, Neurath became a city worker in charge of several such social programs—including housing and visual education.

For publishing activities, the different projects relied on private printing and publishing companies whose operations depended on sales: Meiner and Springer, and later The University of Chicago Press.

[2] See Stadler 2001.

[3] See Beller 1989.

[4] See Stadler 2001 and McFarland et al. 2020. I'm indebted to Friedrich Stadler for a discussion of funding sources.

10.3 Methodological Foundational Debates as a Bridge Between Economy and Philosophy

Neurath's second and more distinctive kind of connection to money was his intellectual, scientific, and administrative engagement with economic matters. Following in his father's footsteps, Neurath's polymathic education at home and the university came to focus on studies in political economy.

In Berlin he pursued doctoral studies with leaders of the Younger German Historical School of Economics, Eduard Meyer and Gustav Schmoller. The result was work on the history of economy in antiquity (1906). Along the way, in Marx's writings Neurath found a model of scientific history of the economy without utopianism, in line with the methodology of German economic historicism.

Neurath also pursued post-doctoral studies with economists associated with Carl Menger's Austrian School of economics: Friedrich Wieser and Eugen Böhm-Bawerk. He became familiar also with the tradition of social utopianism developed by Ballod-Atlanticus and Josef Popper-Lynkeus.

The contrasts between the different approaches fostered a meta-scientific perspective on the social sciences. Economics was a busy arena of debates over goals, methods, and scientific unity. His positions in these debates found their way into the debates in logical empiricism.

10.4 War, Peace, and the Planned Economy

One historical and theoretical possibility intriguing to Neurath was the prospect of a war economy. Neurath drew attention to examples of war economies in antiquity—for instance in Egypt—noting their evolution from monetary exchange and a profit motive to organized barter and a productivity motive. He became acquainted also with economic studies of past wars, including the Napoleonic Wars and the American Civil War, that formed a minor tradition within political economy challenged by rising attention to the economic role of private individual agents.

But political events in Europe such as the Balkan wars (1912–13) and the First World War were to provide him with a new historical experiment and the possibility of a scientific study of modern cases. His critical, empirical, and theoretical insight was, in the spirit of his father's critique of the actual forms of market capitalism, that war economies point to inefficiencies in peacetime market economies.

The conclusion had radical methodological and disciplinary consequences. It compelled many to reject economics conceived of as a price theory of exchange value, especially with cardinal measures of utility, monetary calculation, and the profit motive. Having recourse to history, he endorsed the alternative of an Aristotelian theoretical economics as the scientific study of collective wealth.[5]

[5] Neurath 1911/2004, 1916b/2004 and 1917/2004.

Economics, properly considered from the perspective of war, is the theory of multi-dimensional collective welfare and efficient productivity through the full utilization of all available resources.

Neurath proposed the new field of war economics beyond the parameters set by the study of market capitalism. Neurath proposed also a peacetime model that revolved around a distinctive feature of a war economy: a moneyless economy, a so-called natural or administrative economy, or economy in kind, depending on the emphasis. In practice, the use of money is reduced to, at best, an internal system of accounting, not exchange. Straddling the disjunctions faced by Marxists, utopianists and the Austrian School, Neurath chose a model of moneyless economy based on a subjective theory of utility, close to the domain of empirical sensations.

During the First World War, Neurath served in the Ministry of Defense organizing army provisions and on a commission to study the economy during the war. Notably, so too did Ludwig von Mises. In his first postwar theoretical application, Neurath helped develop a plan for the socialization of Saxony—the so-called "Kranold-Neurath-Schumann plan." Next, during the short-lived Bavarian Soviet Republic (1918–19), he worked for the Office for Central Planning, for which he was tried and briefly jailed and deported back to Austria.[6] By 1919 Neurath had chosen to lead a life of action at the service of the ideal of a planned, democratic, socialist society.

The nature of economic planning in terms of collective wellbeing and efficient allocation of resources had methodological implications for the theory and practice of economics. It raised the issue of the unity of the economic science and the issue of the rationality of decision and governance. These issues would reappear in the context of debates over logical empiricism and Neurath's views on political economics evinced the distinctive character of his contributions to those debates.

Three key dimensions were unity, control and rational decision-making.

Planning for social and economic change required decision-making and was aimed ultimately at control, a political not a scientific category. Neurath identified as central to Marx's critique of capitalism the notion of chaos or anarchy of production. The capitalist order based on money and profit placed, Marx declared in *Capital* and elsewhere, social and economic relations "beyond the control of human agents."[7] In multiple writings, Neurath formulated his thoughts on a planned, administered, moneyless economy in the language of social or state control—of money, production, distribution, and the whole economic life. He wrote in the same terms about the purpose of his proposed plan for the socialization of the economy in Saxony in 1919: "From the beginning we have stood for a complete control,

[6]As a result, he lost his first and last university job after having received his habilitation at Heidelberg in 1917 under Max Weber and the neo-Kantian Heinrich Rickert. Carnap would borrow from Rickert the notion of epistemology as the account of the relation of chaos to knowledge; Neurath followed, by contrast, Marx's notion of the relation of chaos, associated with capitalism, to economic order.

[7]Quoted in Neurath 1925a/2004, 438. Neurath quotes also Engels from *Anti-Dühring* about anarchy and the need to control society as whole on ibid., 440.

embracing production and consumption, for a complete socialization."[8] The different plans similarly included an administrative structure of control councils. Control requires and provides cognitive access, unlike the market mechanism defended later by Hayek: "Everything becomes transparent and controllable."[9] Language entered the picture in the same terms: "To call something by name means to gain control over it."[10]

The theme of unity reappeared in the demand that comparisons and interventions must be unitary, holistic, looking at the whole economy and society.

According to Neurath, a holistic perspective based on comparisons of predicted consequences for the whole was required to engineer the economy. Thus, he wrote that "one can only speak of socialization of the economy as a whole," and that "successful socialization is possible only of the whole and from above."[11] Neurath spoke of a universal statistics concerning complex sets of qualitative and quantitative indicators to represent the irreducible plurality of pleasures. In different social arrangements, different indicators might be correlated, or the same indicators might be correlated differently. Instead of monetary measures of wealth, particular multidimensional combinations of objective measures of life situations and quality form particular (physiognomic) profiles or (topographical) maps.[12]

The key feature of Neurath's proposals is the claim that no uniform commensurating unit of comparison and calculation such as money, labor, or time units is available. Neurath specifically challenged the applicability of quantitative monetary comparisons: "How could a quantity of electricity which a river provides us with be entered as an increase in amounts of labor units?"[13] He allowed, however, that monetary calculations may still be used for accounting purposes.[14]

As Elisabeth Nemeth noted, unity instead of a measurement unit. The task of a coordinated, unified effort at the point of social action appeared symbolized in a discussion of war economics of 1913 by the famous image of a boat which needs to be repaired while afloat at sea symbolizing the lack of absolute epistemic grounding and the irreducible contextuality of decisions making.[15] Next, lamenting the failure of the Bavarian Revolution, Neurath noted: "A unified program could have

[8] Neurath 1919b/1973, 20.

[9] Neurath 1919a/1973, 142.

[10] Neurath 1921/1973, 210.

[11] Neurath 1919/2004. 137, 150.

[12] See Neurath 1937/2004. Representations of the whole included life orders as functions from objective life conditions to subjective life qualities. These would be observable, discrete conditions of life (housing, food, clothing, working hours, etc.) that operationalize the measure of life quality or well-being better than utility.

[13] Neurath 1925b/2004, 468.

[14] Kautsky criticized Neurath for confusing money as capital and money as means of exchange.

[15] See Neurath 1913.

coordinated and unified action."[16] The theme remained in his social thought as the question of "unity of action ... without any kind of authoritative integration."[17]

Planning was justified as rational. But rational planning couldn't be reduced to rational calculation, a fantasy of complete insight he dismissed in 1916 as pseudo-rationalism. Instead of aggregating from individual, precise judgments based on quantitative and universal units of comparison, Neurath declared the unavoidability (and desirability) of imprecise judgment, auxiliary assumptions, and a unified perspective: "From given means a school or a hospital could be built and it was not net profit that decided in favour of one or the other, but judgement, undoubtedly imprecise in many respects, about the expected advance in education with all its consequences on the one hand and, on the other, the gain in health."[18]

Like unity, the philosophical theme of rationality emerged in Neurath's work prior to the logical empiricist debates in a theoretical debate over money. He had joined what he considered a socialist Enlightenment project and in in planning theory he faced the problem of justifying economic decisions. Neurath next attempted to respond to the theoretically oriented critiques of socialist rational planning as promulgated by Ludwig von Mises, Max Weber, and Friedrich von Hayek.

Against socialism, Mises and Weber raised the "rationality problem"—as well as an objectivity problem—which questioned the formal mechanical rationality of economic decisions and decision making. Hayek subsequently raised the "information problem" which posed the practical challenge of finding and processing statistical data needed to make economic decisions rational.

For Mises, economic rationality in capitalism is based on the determination by individuals of the highest profit among choices available to them. For Neurath, by contrast, economic rationality in socialized economies is based on determining the greatest use value to a population as a whole.

More generally, the issue between the critics of socialism and Neurath was the nature of institutional solutions to the coordination problem in a pluralistic society with dispersed local, future, and practical knowledge. For instance, Neurath had defended a democratic, discursive solution of a shared neutral language for social coordination of decision and action whereas by contrast Hayek proposed a non-discursive market pricing mechanism.

The theme of rational coordination, or unity, will connect, on the one hand, money (including moneyless theory) and unified planning to, on the other hand, pictorial linguistic planning (ISOTYPE), architectural and urban unified planning, and the intellectual and scientific unified planning of linguistic, physicalist scientific

[16] Neurath 1919b/1973, 19.

[17] Neurath 1946/1983, 230.

[18] Quoted in Uebel 2007b, 44. Neurath aligned himself with the Austrian disjunctive economic positions: either a free market economy with money or central planning without money. Either a subjective or an objective (Marx, Kautsky and, others' labor theory) theory of value. Menger, Böhm-Bawerk, and others chose a theory of market economy based subjective theory of utility. Neurath chose a moneyless economy based on subjective theory of utility as pleasure and displeasure See also Uebel 2004 and 2007a.

philosophy. In particular, he explained, "for a democratic society, it is important to have a common knowledge in a common language"[19] and "new clarity and purposefulness in communication ... may be regarded as preparation for more incisive social planning."[20]

Here one may follow Antonia Soulez in drawing a distinction between Neurath's and Wittgenstein's respective philosophy of culture as expressed by their views on language and urban life.[21] Each represents a different view of modernity and the exercise of the will. Soulez contrasts Wittgenstein's ethical stoic action on oneself and Neurath's will and plan to change the world and build the future.

10.5 Robinson Crusoe in Political Economy: Money and Language

In the economic writings of Marx, as well as in the debates about Marx's writings which ensued, Neurath encountered doctrines, problems, and perspectives illustrated with references to Robinson Crusoe, the fictional figure in Daniel Dafoe's 1719 novel. The novel generated a genre of so-called Robinsonades which explore situations of individual survival and the individual recreation of civilization using technology but not social collaboration. In the human sciences, the Robinsonades was a trope used to showcase individual capabilities or limitations and ultimately shed light on the necessity or dispensability of social functions.

Marx was one such figure who employed Robinsonades to contrast the modern capitalist situation to that of primitive economies depicted in history or fiction. Cited by Neurath, in the first volume of *Capital* Marx used the trope of Crusoe to critique the wasteful capitalist misallocation of resources in terms of goods and labor power. Crusoe's economic evaluations and decisions are, according to Marx, direct, transparent, simple, and rational; they are social without society.

Marx also considered money a constitutive language that measured and enabled valuation emerging in relations of exchange.[22] Language, money, and labor are social but anchored in individual powers and circumstances exemplified by Crusoe's case.

Wieser and Böhm-Bawerk, in 1891, and Mises, in 1920, also introduced Crusoe in debates about scientific methodology. In Wieser and Böhm-Bawerk, Robinsonades illustrated a defense of a subjectivist conception of value in individual marginal utility theory focused on the exchange value of goods and services. Mises referred to Crusoe in his critique of the notion of moneyless rational planning. Unlike Marx,

---

[19] In Neurath 1996, 256.

[20] Neurath 1933/1973, 223.

[21] See Soulez 1996. Soulez criticizes as simplistic Massimo Cacciari's application of the distinction between *oikos* (Wittgenstein) and city (Neurath) in Cacciari 1993.

[22] See Eich 2022, 105–38; Neurath 1996, 331.

for Mises Crusoe's isolated activities were equivalent to domestic and collectivist perspectives.

10.6 Logical Empiricism: Rationality, Objectivity and Unity

The question of rationality would be later raised and addressed by Carnap and other logical empiricists. It was linked to the role of formal rules in the idealized reconstruction of scientific knowledge and methods considered as a unified, rational, and objective enterprise.

Concerned with rational economic decisions, in 1916 Neurath had put forward a general critique of Cartesian rationalism as a standard for theoretical and practical reasoning.[23] He dismissed such an approach as pseudo-rationalism what he characterized as the failure to admit limits both of actual insight and in the resulting application of the rules of reason for action and belief. The limits of knowledge and the power of reasoning require, instead, the appeal to what he called, equivalently, provisional rules, extra-logical factors, or auxiliary motives.[24]

In the 1930s, Neurath appealed to this view to engage discussions in the Vienna Circle concerning the role for logical tools across the sciences and the rational, logical methods of empirical reasoning.[25] In applying methods of verification and falsification, he stressed the inevitable underdetermination of hypotheses by data and the valuable role of pragmatic considerations. A special target of his critique was Popper and his dogmatic deductivism about falsification.[26]

Already a member of the Vienna Circle, and still in Red Vienna, Neurath continued defending social and economic planning through a controlled economy, for instance in his contribution to the World Social Economic Congress in Amsterdam in August 1931: It is necessary, and possible, "to have direct and central control of the economy."[27]

The formulation of economic plans and critique of capitalism extended now to scientific language and theories with the concern over "uncontrollable elements."[28] In the case of the role of logic, the Circle's manifesto explained in 1929 that "formalizing the intuitive process of inference of ordinary thought" is "to bring it into a rigorous automatically controlled form by means of a symbolic mechanism."[29] In October of the same year, Neurath presented the Circle's so-called scientific world

[23] See Neurath 1916a/1983.

[24] Besides the historical role of instinct and authority, Neurath mentions drawing lots as a pure auxiliary motive; he defends its use as a modern rational way to make a decision when making any decision is better than making none.

[25] See Neurath 1935/1983, and 1936a, b/1983.

[26] See Neurath 1935/1983; Cat 1995.

[27] Neurath 1931b/2004, 503.

[28] Neurath 1996, 322.

[29] Carnap et al. 1929/1973, 309.

conception in a Sunday supplement of the *Workers' Paper*. He pointed to the other mechanism of control over acceptable theorizing, i.e., "control by experience": Logical and, especially, empirical control are the marks of scientific inquiry and the tool for demarcating unified science from the uncontrolled and isolated statements of metaphysics.[30] He referred to protocol statements accordingly, as control statements. He spoke similarly about planning of language, science and unity. Entangled in this development was the importance and interpretation he gave democracy and the role of experts.

Empirical, control statements had to meet the same standard of public testability and non-dogmatic revisability. Empiricism was now physicalism, unifying in line with socialist materialism: "what matters is that statements contain references to the spatio-temporal order."[31]

In his classic paper on protocol statements, the image of the boat of the planned economy he had introduced first in 1913 reappeared as an image of physicalist unified science without firm empirical foundations: "There is no way to establish fully secured, neat protocol statements as starting points of the sciences. There is no *tabula rasa*. We are like sailors who have to rebuild their ship on the open sea, without ever being able to dismantle it in drydock and reconstruct it from the best components."[32]

Similarly, he argued against the availability of a meaningful private language:

> The universal jargon ... is the same for the child and for the adult. It is the same for a Robinson Crusoe as for a human society. If Robinson wants to join what is in a protocol of yesterday with what is in his protocol today, that is, if he wants to make use of language at all, he must make use of the 'intersubjective' language. The Robinson of yesterday and the Robinson of today stand precisely in the same relation in which Robinson stands to Friday .... In other words, every language as such is intersubjective.[33]

In the same period, he would remind readers, writing on rational economic theory, that 'quite a few political economists advocate the thesis that a Robinson Crusoe—or what amounts to the same thing, a controlled economy—calculates in terms of profits and losses.'[34] And he reintroduced the economic version of the boat.[35]

By the 1910s, Neurath had also faced the related questions of unity of the natural and human sciences and unity of socialization plans for moneyless economies. From the late 1920s onward, he adapted the concepts, values, and methods developed in reflection on political economy to the Vienna Circle's "unity of science against metaphysics" as a tool for intellectual and social reform.

[30] Neurath 1936a/1983, 133.

[31] Neurath 1931a/1983, 54. For a discussion of the inception of physicalism and Neider's role see Uebel 2007b, 128–37.

[32] Neurath 1932/1983, 92.

[33] Neurath 1932/1983, 96.

[34] Neurath 1935/1987, 95.

[35] Ibid., 108.

His notion of unity in science and in theory more generally borrowed from his earlier historical methodology and economic planning. He advocated models of information he called encyclopedic, that were open, future-oriented, planned, socially constructed, collaborative, utopian and shared. He stood against the ideal of a closed system of unique, precise formulas. Maritime images of boats and archipelagos substituted for the traditional image of scientific knowledge. Neurath defended the open, encyclopedic model of political and scientific orchestration rather than scientific or state hierarchical pyramidism.[36]

10.7 Wittgenstein's Post-*Tractatus* Philosophy of Language: Money Enters the Picture

Ludwig Wittgenstein was born into an educated and wealthy Viennese family of assimilated Jewish background. His father, Karl, was a steel magnate and art patron. By the end of the nineteenth century, Karl held a virtual monopoly on steel in Central Europe as the majority owner of numerous large steelworks and coal and iron mining companies, setting up and controlling Austrian railway and iron cartels and sitting on the board of major banks.[37] Karl Wittgenstein was also an informed analyst of national and international economies and author of multiple polemics. His views were supported by detailed statistics and comparative psychological and social analysis of workers, capitalist entrepreneurs and national policies. Preoccupied by business cycles and prices, he defended social welfare, tariffs and cartels, against neoclassical economics, but was less of an idealist and critical of capitalism than Neurath's father.

Upon Karl's death in 1913, the estate was divided equally between his wife and six surviving children. As a result, Ludwig Wittgenstein became one of the wealthiest men in Europe. After an almost seven year hiatus from academic life, during which he lived as a rural schoolteacher, Wittgenstein returned to Vienna in 1926. In Vienna he was intellectually reenergized by new debates in the foundations of mathematics and by the sometimes critical discussions of the *Tractatus* by members of the Vienna Circle, especially Schlick and Waismann. Neurath led the charge of antimetaphysical denunciations. Then in 1929 he accepted a fellowship to return to Cambridge. Since his previous time at Cambridge, Wittgenstein had become interested in physiognomy and anthropology, from which he borrowed organizing concepts and illustrations.[38] Key concepts featured in *Philosophical Investigations*

[36] See Neurath 1943/2004 and 1945/2004; Cartwright et al. 1996, O'Neill 2007. In general, for Neurath, a planned economy is not a technocracy, or tyranny of experts. Nor is it socialist planning—a particular mode of distribution of resources—much less Soviet planning; see Chaloupek 2007. For Neurath, social planning serves democracy.

[37] Nyíri and McGuinness 1987.

[38] On his British anthropology sources, see his critical discussion of James George Frazer's *Golden Bough* in Wittgenstein, *Remarks*.

would be those of family resemblance, language game, and form of life. Under the sway of the concepts, the meaning of language became for Wittgenstein detached from atomic states of affairs and logical form and was, instead, attached to concrete practices; meaning, Wittgenstein declared, is use. Like Neurath, for Wittgenstein the theory and form of philosophy derived from the concrete world of practice and social life.

The social perspective didn't eliminate the focus on individuals. The therapeutic value of philosophizing concerned individual philosophical problems or confusions in the individual thinking mind, not changing the world. Yet, he considered them now the effect and expressions of disorder in the social language games and the way of life of his community. In his struggle with language and society, his was, unlike Neurath's, a fighting but not a reformist attitude.[39] Still, philosophy is something each man must do for himself, and is an ethical duty. As such, it was for Wittgenstein as it was for Descartes in the disciplined method of the cogito's exercise, a matter of the will and not of intellect.[40]

Neurath, too, was interested in both physiognomic and anthropological perspectives on human individuals and societies and used terminology and examples in his early economic works as well as later "philosophical" ones.[41] The 1920s marked a transition in Wittgenstein's outlook from practicing private architecture and disposing of private money to talking about money and to economists. But in contrast to Neurath who insisted on social reform, Wittgenstein concentrated on individual self-therapy.

For Wittgenstein architectural and economic planning were personal family matters.[42] He designed for himself an isolated cabin on a Norwegian fiord (1913). He was also personally involved in constructing a house in Vienna for one sister, Margarete Stonborough (1928) in collaboration with his longtime Viennese friend, the architect Paul Engelmann. After the war, he donated his inheritance to his surviving siblings, except for Margarete. Financially, he relied on school teaching in Austria and university funding at Cambridge. Money became a regular topic of his correspondence with siblings and colleagues.[43]

Like Neurath, Wittgenstein's relation to money and the economy included economic theory, or at least economists. We can only speculate about his acquaintance with his father's views. Back at Cambridge in 1929 he started holding regular meetings with his undergraduate friend Maynard Keynes. He also had frequent discussions with Keynes' Italian economist colleague, Piero Sraffa (an ally critic of

[39] von Wright 1982, 113.

[40] Kenny 1982, 26.

[41] Also Neurath, interested in anthropology of social groups, had borrowed from Frazer, in his case to link modern science and technology to primordial and earlier magic.

[42] Even city planning is for Wittgenstein a philosophical illustration of personal intellectual challenges and uses. In *Philosophical Investigations* he wrote that our language "can be seen as an ancient city"; Wittgenstein 1953, art. 18, 8. The philosopher finds ways to map it and walk without getting lost. This view differs dramatically and tellingly from Neurath's approach to city planning.

[43] See McGuiness 2008, 2018.

neoclassical economics) and Keynes' (and Neurath's) theoretical nemesis, Friedrich von Hayek.[44]

Hayek and Wittgenstein had first met on a train to the Italian Front in the late summer of 1918.[45] After the war, Hayek became familiar with the *Tractatus* and its significance through his mentor Ludwig von Mises and Ludwig's mathematician brother and early Vienna Circle member, Richard. Wittgenstein and Hayek continued meeting at least through the 1940s in Cambridge at the Moral Science Club. In fact, Hayek began even writing about Wittgenstein in 1953 and embarked on a biographical project he never brought to fruition.

Sraffa was acknowledged in the preface to the *Philosophical Investigations*: "I am indebted to this stimulus for the most consequential ideas of this book."[46]

As Wittgenstein was overcoming the *Tractatus*' logical atomism about propositions and facts in the world, Sraffa was developing a critique of the atomistic and individualistic perspective in neoclassical economics and simplistic causal explanations. Sraffa was another critic of neoclassical economics, its atomism of agents' subjective decisions, marginal utility calculations, and simple causal mechanisms explaining prices by appealing to labor, utility, or supply and demand. Prices, for Sraffa, do not carry the information that, according to Hayek, causes individual agents to restore market equilibrium between supply and demand.

Wittgenstein also discussed with Sraffa his own philosophical ideas, although difficulties of communication arose between them.[47] Wittgenstein's student Norman Malcolm provided in his recollections another hint about the developments in Wittgenstein's philosophical outlook:

> Wittgenstein was insisting that a proposition and what it describes must have the same 'logical form,' the same 'logical multiplicity.' Sraffa made a gesture, familiar to Neapolitans as meaning something like disgust or contempt, of brushing the underneath of his chin with an outward sweep of the finger-tips of one hand. And he asked: 'What is the logical form of that?'[48]

[44] Crudely put, Keynes defended of government fiscal and monetary intervention, notably in *Treatise on Money* and *The General Theory of Employment, Interest and Money*. Hayek defended free markets in *Prices and Production* and *Collectivist Economic Planning* against rational socialist planning, following von Mises' views on the economic calculation problem. According to Hayek, changing prices communicate local and individual information, price signaling in markets as information-knowledge machines. Wittgenstein had participated in discussions with Keynes, Russell and others in discussions of probabilities prior to the publication in 1921 of the *Tractatus* and Keynes' Treatise on Probability (for details see Matthew Coates' forthcoming HOPOS article 'Keynes, Wittgenstein, and Probability in the *Tractatus*').

[45] Quoted in Erbacher 2019, 9.

[46] Wittgenstein 1953, x.

[47] Two letters to Sraffa of 1934 have been published with commentary in Iacco 2019. See also especially letters from 1932, 1934, and 1935 in McGuiness 2008, 196, 222–29, 235, and 248.

[48] Malcolm 2021, 57–58.

10.8 Money in Philosophy of Language

Wittgenstein drew upon the case of money in his post-*Tractatus* writings about language. His examples conjured up conceptual possibilities operating as philosophical machines—much like Neurath's (social) engineer's possible machines informing social plans and utopias. Unlike Neurath or Marx, Wittgenstein did not consider the relation between language and money as part of a social theory. Instead, he used references to money rhetorically, as analogies for the purpose of philosophical illustration and evidence.

Wittgenstein's money references describe individual use and talk of money as part of a language game and a social practice. Following rules and concrete practices within social or institutional settings were forms of individual behavior that interested Wittgenstein, especially when they concerned language and mathematics. Several passages from the *Philosophical Investigations* draw on the case of money in order to demonstrate different features of linguistic meaning. For instance[49]:

> You say: the point isn't the word, but its meaning, and you think of the meaning as a thing of the same kind as the word, though also different from the word. Here the word, there the meaning. The money, and the cow that you can buy with it. (But contrast: money, and its use.)[50]

Another passage presents Wittgenstein's well-known argument against the notion of private language, in the wake a similar critique by Neurath:

> Why can't my right hand give my left hand money?—My right hand can put it into my left hand. My right hand can write a deed of gift and my left hand a receipt.—But the further practical consequences would not be those of a gift. When the left hand has taken the money from the right, etc., we shall ask: "Well, and what of it?" And the same could be asked if a person had given himself a private definition of a word; I mean, if he has said the word to himself and at the same time has directed his attention to a sensation.[51]

As illustrated through analogies with money, the use of language relies on social institutions and rule following. Notice the similarity both to Neurath's Crusoe argument and to the earlier economic readings of Crusoe. Both linguistic and economic activities are radically social. In Neurath's argument for physicalism, the Crusoe example motivated primarily the social warrant for the constancy of language use in the scientists' cognitive life.

Wittgenstein was concerned to undermine the notion of a private language in the sense of words whose meaning is essentially dependent on the inner experience or individual consciousness of a subject. Wittgenstein followed Neurath in arguing that it is impossible—practically for Neurath, logically for Wittgenstein—for an individual to use consistently, and for others to understand, a private language. In a note written in 1929, Wittgenstein referred to the primacy of our public everyday

---

[49] Other examples appeared in, among other works, his *Philosophical Grammar* and *Zettel*.

[50] Wittgenstein 1953, art. 120.

[51] Wittgenstein 1953, art. 268.

language about physical things over a language of immediate sensations. In the *Investigations* he went so far as to say that language is dependent on the public social institutions of language-games and their rules.

A private language is, by contrast, impossible because its use requires following its rules, which in turn requires a private modality of rule-following, and this, Wittgenstein claims, is incoherent. That is because to follow a rule privately means that thinking one is following a rule is the same as actually following it.[52] But this assumption is based on a subjective internal notion of justification, dependent on intrinsic individual consciousness. And this is incoherent, for according to Wittgenstein, "justification consists in appealing to something independent."[53] The independence is provided by a notion of language use and rule following with reference, actual or potential, to membership in a wider external community and to public or social interaction.

The philosopher Saul Kripke and the sociologist David Bloor have pointed out that this condition of normative constraint by intersubjective dependence is not incompatible with Robinson Crusoe's physical isolation.[54] In fact Kripke and Bloor even draw an analogy between the impossibility of a private language and the impossibility of central planning as it features in the thought of Ludwig von Mises. According to von Mises, if an economic agent has no extrinsic range of intersubjective and commensurable values—that is, the numerical market prices set independently by others—then he has no rational and objective justification for choosing between alternative courses of action. For von Mises exact numerical commensurability and the rules of explicit calculation are the basis for both objectivity and rationality.

10.9 Money in Philosophy of Mathematics

Wittgenstein's philosophy of mathematics is constructivist and finitist. That is to say, mathematics is invented, not discovered; and, like philosophy, it is an activity, a practice. There are no mathematical facts (extensions) or meaningful propositions, only finite procedures of calculation, including counting and proof. As in the case of language, mathematics was also for Wittgenstein a matter of concrete practices in a social setting. Counting was one such practice, and counting money was a recurrent example addressing the question of the reliability of counting procedures such as using one-to-one correlations and comparisons. From an epistemic, representational standpoint, two related problems arise: the problem of regress and the problem of assessing discrepancies. In reaction to the logicism of his teachers Frege and

---

[52] Wittgenstein 1953, art. 202.

[53] Wittgenstein 1953, art. 265.

[54] See Bloor 1997 and Kripke 1982.

Russell, Wittgenstein claimed that arithmetic is not a matter of logic alone but depends on the practices in which that logic is exercised:

> And if we put stacks of shillings side by side, they'd correspond one-one. This is so. But it needn't be so.—What if we don't get the same result? We'd say, "I've made a mistake" or "A shilling has vanished," etc. ... But it is not clear what we should accept as clearing up the mistake, why we should take this as reliable and the other not.[55]

In order to count, to calculate a total amount and assign it the correct symbol, we decompose a quantity into units and use a notation of ordered numerals to label the units in increasing order according to a one-to-one correlation, in a procedure we can survey and apply reliably. Counting money works precisely in this way.[56]

Similarly, according to Wittgenstein knowing the rules of mathematics means knowing how to correctly apply the rules of mathematics, and that it is a matter of our accepted standards:

> Imagine people who used money in transactions; that is to say coins, looking like our coins, which are made of gold and silver and stamped and are also handed over for goods—but each person gives just as what he pleases for the goods, and the merchant does not give the customer more or less according to what he pays. In short this money, or what looks like money, has among them a quite different role from among us.[57]

In these comments from *Remarks on the Foundations of Mathematics* Wittgenstein again reverts to a habit acquired in *Philosophical Investigations* and other post-Tractarian works of using money and coinage metaphors to illustrate and elaborate wider philosophical points about the nature of language, meaning, and logical validity.

10.10 Conclusion

The cases of Neurath and Wittgenstein are significant examples of the different roles that considerations of money have played in the development of philosophy considered broadly to include the foundations of mathematics and science. Presenting them together sheds light on their shared historical and intellectual sources. It also exhibits shared assumptions about money as an accounting unit, a minimal philosophy of money.

But it throws into sharper relief their specific differences, for instance, Wittgenstein's personal and instrumental use of money and money-talk stands out against Neurath's active, constructive, social, and scientific approach, theoretically rejecting it, linked both to socialist policies and economic theory. Both cases also illustrate varieties of ways in which considerations of money and other relations to

---

[55] Wittgenstein 1939/1976, 257.

[56] Wittgenstein 1956, 155.

[57] Wittgenstein 1956, 95.

it contribute to intellectual work. And this independently of different interpretations of Crusoe's situation and especially of debates over Wittgenstein's claims.

References

Beller, S. 1989. *Vienna and the Jews 1867–1938*. Cambridge: Cambridge University Press.

Bloor, D. 1997. *Wittgenstein, Rules and Institutions*. London: Routledge.

Cacciari, M. 1993. *Architecture and Nihilism*. New Heaven: Yale University Press.

Carnap, R., H. Hahn, and O. Neurath. 1929/1973. *The Scientific Conception of the World: The Vienna Circle*. In O. Neurath 1973, Dordrecht: Reidel, 299–318.

Cartwright, N., J. Cat, K. Fleck, and T.E. Uebel. 1996. *Otto Neurath: Philosophy Between Science and Politics*. Cambridge: Cambridge University Press.

Cat, J. 1995. The Popper-Neurath Debate and Neurath's Attack on Scientific Method. *Studies in History and Philosophy of Science* 26 (1995): 219–250.

———. 2024. Money and Philosophy in Vienna: Otto Neurath and Ludwig Wittgenstein. In *Palgrave Handbook of Philosophy and Money*, ed. J. Tinguely, vol. 2. London: Palgrave.

Chaloupek, G. 2007. Otto Neurath's Concepts of Socialization and Economic Calculation and His Socialist Critics. In *Otto Neurath's Economics in Context*, ed. E. Nemeth, S.W. Schmitz, and T.E. Uebel, 61–75. Vienna: Springer.

Eich, S. 2022. *The Currency of Politics*. Princeton, NJ: Princeton University Press.

Erbacher, Ch., ed. 2019. *Friedrich August von Hayek's Draft Biography of Ludwig Wittgenstein*. Paderborn: Mentis.

Iacco, M. de. 2019. Wittgenstein to Sraffa: Two Newly Discovered Letters from February and March 1934. *Nordic Wittgenstein Review* 8 (2019): 209–223.

Johnston, W.M. 1972. *The Austrian Mind*. Berkeley: University of California Press.

Kenny, A. 1982. Wittgenstein on the Nature of Philosophy. In *Wittgenstein and His Times*, ed. B. McGuinness, 1–26. Oxford: Basil Blackwell.

Kripke, S.A. 1982. *Wittgenstein on Rules and Private Language*. Cambridge, MA: Harvard University Press.

Malcolm, N. 2021. *Ludwig Wittgenstein: A Memoir*. Oxford: Clarendon Press.

McFarland, R., et al., eds. 2020. *The Red Vienna Sourcebook*. Rochester, NY: Camden House.

McGuiness, B., ed. 2008. *Wittgenstein in Cambridge, Letters and Documents 1911–1951*. London: Blackwell.

McGuinness, B., ed. 2018. *Wittgenstein's Family Letters*. London: Bloomsbury.

Neurath, O. 1911/2004. On the Theory of Social Science. In O. Neurath 2004, 265–291.

———. 1913. Problem der Kriegwirtschaftslehre. *Zeitschrift für die gesamte Staatswissenschaft* 69 (1913), 438–501.

———. 1916a/1983. The Lost Wanderers of Descartes and the Auxiliary Motive." In O. Neurath 1983, 1–12.

———. 1916b/2004. Economics in Kind, Calculation in Kind and Their Relation to War Economics. In O. Neurath 2004, 299–311.

———. 1917/2004. The Conceptual Structure of Economic Theory and Its Foundations. In O. Neurath 2004, 312–341.

———. 1919/2004. Economic History in Antiquity [Excerpts]. In O. Neurath 2004, 120–152.

———. 1919a/1973. Character and Course of Socialization. In O. Neurath 1973, 135–150.

———. 1919b/1973. Autobiographical Excerpts from Otto Neurath. In O. Neurath 1973, 18–28.

———. 1921/1973. Anti-Spengler. In O. Neurath 1973, 158–213.

———. 1925a/2004. Economic Plan and Calculation in Kind. In O. Neurath 2004, 405–465. Dordrecht: Kluwer.

———. 1925b/2004. Socialist Utility Calculation and Capitalist Profit Calculation. In O. Neurath 2004, 466–472.
———. 1931a/1983. Physicalism. In O. Neurath 1983, 52–57.
———. 1931b/2004. The Current Growth in Global Productive Capacity. In O. Neurath 2004, 475–505.
———. 1932/1983. Protocol Statements. In O. Neurath 1983, 91–99.
———. 1933/1973. Museums of the Future. In O. Neurath 1973, 218–223.
———. 1935/1983. Pseudorationalism of Falsification. In O. Neurath 1983, 121–131.
———. 1935/1987. What is Meant by a Rational Economic Theory? In *Unified Science*, ed. B. McGuinness, 67–109. Dordrecht: Reidel.
———. 1936a/1983. Individual Sciences, Unified Science, Pseudorationalism. In O. Neurath 1983, 132–138.
———. 1936b/1983. Encyclopedia as 'Model'. In O. Neurath 1983, 145–158.
———. 1937/2004. Inventory of the Standard of Living. In O. Neurath 2004, 513–526.
———. 1943/2004. Late Reflections on the Theory of Planning. In O. Neurath 2004, 527–538.
———. 1945/2004. Alternatives to Market Competition. In O. Neurath 2004, 546–548.
———. 1946/1983. The Orchestration of the Sciences by the Encyclopedism of Logical Empiricism. In O. Neurath 1983, 230–242.
———. 1973. In *Empiricism and Sociology*, ed. M. Neurath and R.S. Cohen. Dordrecht: Reidel.
———. 1983. In *Philosophical Papers 1913–1946*, ed. R.S. Cohen and M. Neurath. Dordrecht: Reidel.
———. 1996. Visual Education. In *Encyclopedia and Utopia*, ed. E. Nemeth and F. Stadler, 245–335. Dordrecht: Kluwer.
———. 2004. In *Economic Writings*, ed. T.E. Uebel. Dordrecht: Kluwer.
Nyíri, J.C., and B. McGuinness. 1987. Introduction. In *Politico-Economic Writings* by Karl Wittgenstein, J.C. Nyíri ed. Amsterdam/Philadelphia: John Benjamins, Publishing Company, i–xl.
O'Neill, J.F. 2007. Pluralism and Economic Institutions. In *Otto Neurath's Economics in Context*, ed. E. Nemeth, S.W. Schmitz, and T.E. Uebel, 77–100. Vienna: Springer.
Soulez, A. 1996. Otto Neurath and the Will to Plan. In *Encyclopedia and Utopia*, ed. E. Nemeth and F. Stadler, 221–232. Dordrecht: Kluwer.
Stadler, F. 2001. *The Vienna Circle: Studies in the Origins, Development, and Influence of Logical Empiricism*. Vienna: Springer.
Uebel, T.E. 2004. *Introduction: Neurath's Economics in Critical Context*. In O. Neurath 2004, 1–108.
———. 2007a. Otto Neurath as an Austrian Economist: Behind the Scenes of the Early Socialist Calculation Debate. In *Otto Neurath's Economics in Context*, ed. E. Nemeth, S.W. Schmitz, and T.E. Uebel, 37–60. Vienna: Springer.
———. 2007b. *Empiricism at the Crossroads*. Chicago: Open Court.
Wittgenstein, L. 1939/1976. *Wittgenstein's Lectures on the Foundations of Mathematics*, Cambridge, 1939, Edited by Cora Diamond, from the notes of R.G. Bosanquet, Norman Malcolm, Rush Rhees, and Yorick Smythies. Ithaca, NY: Cornell University Press.
———. 1953. *Philosophical Investigations*. Trans. G.E.M. Anscombe. Cambridge: The Macmillan Company.
———. 1956. *Remarks on the Foundations of Mathematics*. Trans. G.E.M. Anscombe, G.H. von Wright, and Rush Rhees. London: Blackwell.
Wright, G.H. von. 1982. Wittgenstein in Relation to His Times. In *Wittgenstein and His Times*, ed. B. McGuinness, 108–120. Oxford: Basil Blackwell.

Chapter 11
Susan Stebbing on the Scientific Attitude and Moral Philosophy Context and Comments

Silke Körber and Adam Tamas Tuboly

11.1 L. Susan Stebbing and Otto Neurath: Setting the Scene

In recent years, there has been an almost exponentially growing tendency to study and evaluate the life and work of L. Susan Stebbing.[1] She was the first female professor of philosophy in Britain, and thus could serve as a role model for many inquiries of feminist studies within the history of analytic philosophy (for example in Janssen-Lauret 2024). Stebbing was not just a recognized and respected formal logician, but represented an early, socially and politically engaged form of analytic philosophy. Her papers, books, and pamphlets were formed in and shaped by current public events; she was sensitive to debates of the time and never hesitated to take an active role in helping others, be they scholarly refugees, young children, or the (un)educated masses that confronted on a daily basis the latest and fallacious mass-communication and propaganda machineries.[2] Her practical logic (see, for example, Stebbing 1934) was meant to serve a debunking role within public policy

[1] See, for example, Beaney (2003), Chapman (2013), Beaney and Chapman (2017/2021), Janssen-Lauret (2017, 2022), Tuboly (2020), Coliva (2021), Douglas and Nassim (2021), Pickel (2022), and Moravec and West (2023).

[2] Stebbing's most detailed biography is Chapman (2013), where most of her public engagements are detailed as well, including the girls' school that she directed.

S. Körber (✉)
Independent Scholar, Humboldt Universität zu Berlin, Berlin, Germany
e-mail: silke.koerber@alumni.hu-berlin.de

A. T. Tuboly (✉)
MTA Lendület Values and Science Research Group, Research Centre for the Humanities, Budapest, Hungary

Department of Behavioural Sciences, Medical School, University of Pécs, Pécs, Hungary

G. Schiemer (ed.), *The Legacy of the Vienna Circle*, Vienna Circle Institute Yearbook 30, https://doi.org/10.1007/978-3-031-80568-4_11

and everyday life. Thus, Stebbing was a many-faceted philosopher, logician, and public intellectual, whose ideas are still relevant in debates about the place and future of philosophy in demented times.

Stebbing was also at the forefront of the internationalization of the Vienna Circle during the 1930s, when she corresponded with leading figures such as Moritz Schlick, Rudolf Carnap, and especially Otto Neurath. She either invited them to England or reviewed their works. She cooperated with them on various projects by joining the editorial boards of Neurath's initiatives and co-organizing the *Fourth International Congress for the Unity of Science* at Girton College, Cambridge (dealing with the language and logic of science).[3] Quite obviously, Neurath had a special relationship with Stebbing: they also invited each other regularly to England and to the Netherlands (where Neurath was living between 1934 and 1940), and it was Stebbing who did the most to foster the release of Neurath and Marie Reidemeister from the internment camp where they were deported as "enemy aliens" after they escaped from The Hague on a small boat. Given that Neurath is still one of the most studied figures of the Vienna Circle, both from systematic and historical standpoints, it made sense for us to compare their lives and works.[4]

During our initial take on the Stebbing-Neurath relationship (Körber 2019, 2023; Körber and Tuboly 2025a), we came across two very interesting, but either unknown, or hardly accessible texts of Stebbing. We thought it would be worthwhile to make both texts available for interested readers as they shed some new and interesting light on what occupied these politically and socially engaged scholars during the Second World War, and especially how Stebbing approached ethics, moral philosophy, public policy, the nature and role of philosophy, and the place of science in society. We focus here on Stebbing's texts as Neurath's unpublished papers from this period have been dealt with in a different volume (Burke and Tuboly 2024), and a selection of their correspondence will be published elsewhere (Körber and Tuboly 2025b).

Hence, in this short editorial introduction we provide the context of these papers and point to some interesting and important questions about their contents, but let the community decide about their significance within the emerging Stebbing-scholarship.

[3] She was also critical with different views of the logical empiricists, not sparing any of the factions within the Circle. See Beaney (2003, 2016) and Franco (2023).

[4] About recent studies on Neurath see the chapters in Cat and Tuboly (2019), about Neurath's relevant British years see Burke and Jansen (2022), Körber (2023), Burke and Tuboly (2024), and da Cunha (2024).

11.2 "The Inadequacies of the Scientific Attitude"

11.2.1 Contexts

The "Inadequacies of the Scientific Attitude" was a hand-written, unpublished text that was sent to Otto Neurath. The manuscript is being kept in the Neurath-archive and that is also where we came across it, and used that copy for transcription here. There was no date on the manuscript, but we estimate it to be written around 1941/42 as it concerns a small book, titled *The Scientific Attitude*, written by the famous developmental biologist C. H. Waddington. The book appeared in 1941 as a inexpensive and accessible Penguin paperback in the Pelican series.[5]

The Neurath-Stebbing correspondence ends around February 1941 when Neurath and Marie got out of internment, and there is no indication of Waddington in previous letters, or the fact that Stebbing has sent the manuscript to Neurath. Between February 1941 and September 1943 (when Stebbing passed away), their relationship had progressed from a constant exchange of letters and a few personal meetings to a regular personal contact and that is where they might have discussed questions of the scientific attitude (that was, of course, a recurring topic in their correspondence before). Because of that, there is no indication about the nature or origins of the manuscript. But given the casual style of Stebbing's paper and the occasional outreach to the audience, it could have been a draft for a talk somewhere.

The book put Waddington in the center of public attention, with a general discussion in the pages of *Nature* in the fall of 1941, following Waddington's short paper on "the relation between science and ethics" (Waddington 1941b). Numerous scientists and philosophers participated in the discussion representing various positions and having different angles; among them were the philosopher C. E. M. Joad, the pharmacologist Chauncey D. Leake, the physicist G. Burniston Brown, the chemical physiologist Arthur David Ritchie, the evolutionary biologist Julian S. Huxley, the physicist-philosopher Herbert Dingle, the historian of science Joseph Needham, and many others, including Stebbing (1941a). The debate unfolding in the prestigious journal was so successful and enlightening that the papers, commentaries, and letters to the editor were collected and republished (with some further materials) in a book called *Science and Ethics* (Waddington 1942). In the introduction, Waddington wrote that he was looking for an intellectual basis for ethics, but Wittgenstein told him once that "this is a terrible business just terrible! You can at best stammer when you talk of it." Waddington conceived the book and the discussion as "communal, perhaps even co-operative, stammering" (1942, 7). The book got a second impression in 1944.

[5]A second, revised edition appeared after the Second World War (Waddington 1948). In the new edition's preface, now with all the war and after-war experiences, Waddington stayed confident in his judgment that problems around our life are not technological—"we already have all, or at least most, of the techniques we need to provide a decent civilised life for everybody"—but "spiritual" and thus concerns "ideals and values" (Waddington 1948, vii).

After the debate in the pages of *Nature*, there was a joint meeting of the Cambridge Moral Science Club and the Aristotelian Society devoted again to the relations between science and ethics; the talks of Waddington, A. C. Ewing, and C. D. Broad were published right after that (Waddington, Ewing, Broad Waddington et al. 1941–1942). There is no indication, among the published speeches, however, that Stebbing read her paper there, although the topic and details would have fitted the occasion; and based on the reference to Waddington's *Nature*-article (and such claims about its content as "you will remember" that paper), the manuscript was prepared after the appearance of the book, and was read around that occasion.

On the surface, Waddington propagated socialism, planning of science and economy, and a certain form of value-ladenness of science. It was a short, harsh, to-the-point booklet, a direct statement of a scientist about ethics, science, public policy, and the role of politics in these matters. It touches upon such questions that were on the forefront of the thinking of British Marxist-communist scientists and public intellectuals during the 1930s and 40s.[6]

Waddington started from the problem that it would not suffice to defeat the Nazis physically. What people needed after the war was some sort of intellectual and cultural redirection, something that would convey new ideas, ideals, and ways of life. After finding many suspects and suggestions lacking, his idea was that *(natural) science* could provide a harmonious and consistent way of life, interfering obviously with ethics and having a word in good and evil. For that, one has to see that science is ethically not neutral and is done in a human society with interest-related agents. The idea was not simply that science could solve special technical problems, but that "it has also developed an attitude to the world which makes some things seem valuable and others not", that is, it delivers "a general way of life" (Waddington 1941a, 9, 86). By adopting the scientific attitude (a love for facts, interest in posing questions, and making penetrating criticism), one could tackle all the current 'social chaos', 'disintegration', and non-economic issues of civilization more efficiently.

Science is not interested in the particular, thus, even in the case of society, the scientific results and directions are concerned with the many and not the individual—it is "concerned with the welfare of society as a whole" (Waddington 1941a, 94). To handle such a mass, rational control has to be introduced, that is, it must contribute to the planning of everyday life. As Waddington (1941a, 124) summarizes at the end of the book the gist of his proposal: the new system will be "centralised, and totalitarian in the sense that all aspects of the economic development of large regions are consciously planned as an integrated whole." With an explicit admiration for Russian communism, Waddington obviously confessed to the radical communist extreme of the 1930s (along the lines of Bernal and Haldane, the heroes of his short treatise).

It was thus basically inevitable for Stebbing and Neurath to get into contact with Waddington. Neurath and Waddington met personally at the Terminology Workshop that Neurath organized in Oxford in October 1941 (with Stebbing and Joseph

[6] On some of the issues and their context see Werskey (1978) and Burke and Tuboly (2024).

Lauwerys being co-organizers, see Burke and Tuboly 2024, ch. 6), and Neurath emphasized in their letters that Waddington's "attitude towards the big bulk of problems of co-operation etc is similar to our own", that is of logical empiricism, and that he is looking for Waddington's general advice as he promotes 'a comprehensive scientific attitude' like Neurath.[7] Behind the scenes, however, Neurath was much more critical of Waddington's approach. When Hayek mentioned Waddington's *The Scientific Attitude* in *The Road to Serfdom* as an example of totalitarianism, Neurath commented 'a sad book'; when Hayek wrote that Waddington's book is a good example of how scientists get greater political power, Neurath marked 'technocrats'.[8] Bringing scientists to the forefront without further ado was a socially dangerous move and an epistemologically absolutist, unjustified idea.

Neurath obviously saw in Stebbing a public intellectual, a "friend of the people" who raised her voice against "extravagances of scientists" (like in the case of physicists who propagated a social philosophy of idealism).[9] To keep popularizing scientists, and even politicians under pressure, Neurath thought that one has to show a more faithful and critical "scientific attitude", which is "more or less a question of 'public health'".[10] Stebbing was indeed in the forefront of discussing, analyzing, and criticizing bad practices of the scientific attitude and in forming good examples. Neurath viewed her publications, especially *Thinking to Some Purpose* and *Philosophy and the Physicists*, as providing principles for a "comprehensive criticism" in the political and public arena; Stebbing was "an expert in 'scientific attitude'" for Neurath.[11]

Besides the practical guides (such as *Logic in Practice*, *Thinking to Some Purpose*) and the particular attempt to debunk and criticize the physicists (*Philosophy and the Physicists*), Stebbing made another, unpublished effort to deal with the highly influential Waddington as well—that resulted, after all, in the most explicit and comprehensive (meta-)analyses of the "scientific attitude" and how it differs from the available tendencies of scientists—the text that has been reproduced here.

---

[7] Neurath to C.H. Waddington, October 31, and November 12, 1941 (ONN/319).

[8] Neurath's copy of Hayek's *The Road to Serfdom*, p. 143 (ONN).

[9] On the historical and public aspect of the physicists-issue see Tuboly (2020), on the philosophical stakes see West (2022a, b). Margaret Macdonald (1943, 16), student of Stebbing gave one of the best and succinct description of the book "in which the philosophic wonderland of Jeans and Eddington with its mysterious and familiar tables, its electrons having Free Will and other ancient fallacies in Cavendish fancy dress, is very wittily analysed."

[10] Quotations from Neurath to Stebbing, February 22, 1938 (ONN/303).

[11] Neurath to Stebbing, June 13, 1939, and March 8, 1940 (ONN/304).

11.2.2 The Manuscript

Stebbing called Waddington's book "interesting but in some ways absurd". In the short text, she summarizes the volume's main points and concerns, then criticizes them, and finally shows that the "universal adoption" of the scientific attitude "would not suffice to cure the ills of our world."

It was an important point for Waddington that science does not function as a guide, or as a simple instrument in pointing towards important steps to reach our values and ideals. In fact, science provided the ideal and values themselves. Waddington was not a simple instrumentalist (like Neurath seems to be on certain occasions, being committed to the view that science helps us in understanding the consequences of our decisions), but he thought that science must fulfill greater and more far-reaching tasks. Science does not examine the consequences or draw up long lists of logical options, but tells us what to do and believe, especially in matters of life and values, even in politics.

For Waddington, thus, science had a moral standing and was not ethically neutral. The scientific attitude, stemming from science itself, was characterized by special dualisms: it was "matter of fact", "empirical and unprejudiced", and "objective" rather than being "romantic", "a priori", and "subjective" (Waddington 1941a, 47). Such dualisms were highly objectionable to both Neurath and Stebbing, thus it was important for both of them to develop a more nuanced approach to the scientific attitude.

While Stebbing obviously accepted some form of the "scientific attitude" (just like Neurath, see above), she argued against Waddington's main point, namely that science is ethically not neutral. Her argument involved the following points:

(1) Waddington argued that from the fact that scientists as human beings bring to science not just their intellectual parts and cognition, but their entire human behavior and aspirations, means that science is ethically non-neutral.

Stebbing disagreed. She thought that Waddington conflated scientists with science. The truth, or empirical confirmation of scientific statements does not depend on our personal aspirations and actions. We might have developed sensitivities towards ideas and facts, but in themselves, they are independent from our subjective take on them. Even if certain scientific statements could be used in ethical reasoning, they are neutral as they are; their content does not determine human action (a moral issue). While consequences of scientific discoveries could have moral relevance (e.g., Fritz Haber's chemical discoveries that both have agricultural and war-related effects), "there is nothing in the least ethical, or moral, in the scientific discovery itself. The discovery is *amoral*, or if you prefer, non-moral." The emotional involvement and excitement of individual scientists does not affect the ethical neutrality of the *content* of science. As Stebbing wrote elsewhere,

> Science is not a goddess or a woman. We cannot ask science, but only scientists. Moreover, we must ask our questions of the scientist at a moment when he is in a scientific temper, capable of giving us 'the ascertained facts and provisional hypotheses' without any admix-

ture of the emotional significance which he reads into these facts in his scientific moods. (Stebbing 1937, 9)

She aimed for an instrumental view of science, where "specialist studies" provide people with much of the information that is relevant and required to make deliberate choices and decisions. That is, after making explicit our values, preferences, and wishes, science can help us to find out means to our ends. "But the scientist – in his capacity of a scientist – does not determine these ends."[12]

(2) The second point of Waddington that Stebbing dealt with concerned the ethical connotation of science. Waddington believed—and Stebbing agreed with him up to a certain point—that science can set up high moral standards and make a contribution that would be desirable to everyone in society. Those included intelligence, impartial and dispassionate thinking, submissiveness to facts, and cooperation. "These characteristics are, I should agree, valuable characteristics", wrote Stebbing. "Would that we all had them!"[13]

But Stebbing went on to analyze further Waddington's commitments. If the scientific attitude comes with the idea of having valuable characteristics even in everyday life, then she happily accepted that. But "if it meant that the maintaining of a scientific attitude *entails* an ethical standard, then that seems to be absurd", and according to her interpretation, Waddington went for that stronger commitment. The idea was, for Waddington, that whatever the direction of evolution is, it is good; that direction is revealed, studied and understood by science, thus science is capable now to make an ethical contribution to our life—or so did Stebbing reconstruct Waddington's line of argumentation. She was not convinced, to say the least, and she went on in a quite rhetorical manner to make a point about the dimension of Waddington's mistake:

Until I read Dr. Waddington's writings I had supposed that the fallacies committed by Herbert Spencer had been so often so conclusively refuted that even the most aged of Victorians would not have revived them. I am sorry to find I am mistaken, and that the dead horses of the Victorian era have been resuscitated by a young Cambridge scientist. I suppose that work has to be done all over again. I have time here only to make a few brief comments.

In making some critical arguments about the connection between the direction of evolution and morality, Stebbing did not aim to question the importance of science

[12] Neurath had a more dialectical picture of the sciences, and instead of introducing "wishful thinking" (something he perhaps attributed to such people as Waddington who propagated unfounded views, that is, their wishes), he wrote about "thinkful wishing", a guided and educated imagination that take into account various possibilities and "utopias"; see Neurath (1944, 31, 43).

[13] It has to be noted, though, that Stebbing seems to think that muddled thinking always results in muddled action, thus we have to clarify our cognitive and intellectual habits in order to arrive at the best available democratic and rational practices. But as Eric Schliesser (ms.) has emphasized, "it is not wholly unreasonable [to] claim that clarity of cognition may often (perhaps always) be unnecessary to achieve our most practical aims (because relying on tradition, faith, instinct, testimony, or expertise/authority of others)." Neurath could not agree more.

in shaping our modern world, even on the level of public policy and morality. But she emphasized yet again that we are not dealing, after all, with science here, but with scientists. Whenever science is about to be used outside of particular and specialized fields, it is in fact individual scientists who have a special humanitarian vision and mission, and nothing guarantees that an accumulated amount of expertise and knowledge in one field could be transposed into another without further ado. Expertise is not transitive. Organisms are part of the world, just like societies, but expertise in organisms does not mean expertise in societies. Scientists' expertise can break down at a point:

> '[U]nscientific science' is a contradiction in terms; 'unscientific scientist carries out a scientific experiment' is a contradiction in terms; but 'unscientific Mr. X, who is a scientist, deciding questions of policy in non-scientific matters' is so far from being a contradiction in terms that, I am afraid, he is to be met on the staff of every university.

Stebbing thus agreed with Neurath about the dangers of technocracy and saw similar problems in Waddington's proposal. Science is not the only authority within society, although it has the important task—alongside philosophy—in clarifying all the matters that have to be decided within our public life.

For those decision-making processes, Stebbing thought that science could serve as a blueprint, a guide that helps us to keep track of things. In her papers and books that related to public policy, she introduced values and processes that would have kept the reader and citizen safe from unfounded propaganda, fallacies of argumentation, and misleading persuasion. She was generally in favour of democratic checks and balances. This is basically a view that emphasizes the importance of the values and characteristics of science and especially of the scientists for the values of society.

This idea—in the context of how values and science are interacting—has been recently addressed by Harry Collins (2023, previously by Collins and Evans (2017). As he wrote:

> How can we not want a society that values honesty, readiness to listen to criticism, universality, corroboration and falsifiability, in political and economic judgements as well as scientific judgements – these are what I'll call 'the good values'? If you would prefer another set of values to be dominant in society, such as meaningless slogans instead of honesty, unwillingness to listen to criticism, prejudice against persons with certain physical characteristics, unwillingness to think that corroboration of a claim is more important than the person who makes it, and unwillingness to countenance the possibility that ideas are better if they are potentially defeasible by evidence – what I'll call 'the bad values' – then this paper will have no appeal. (Collins 2023, 272–73)

Stebbing could be seen thus as an early proponent of scientific values influencing public values (or characteristics), and thus the importance of her views for all our contemporary challenges (stemming from science-denial, skepticism, conspiracy, and populism), could have a renewed interest and importance.

11.3 Men and Moral Principles

11.3.1 Contexts

Stebbing's posthumously published booklet, *Men and Moral Principles*, was also sent to Neurath, who kept a copy in his library, and made a few marginal notes—making evident his interest and engagement with Stebbing's ideas and work.[14]

Stebbing's text was published in 1944 as the 13th L. T. Hobhouse Memorial Trust lecture, a series that was dedicated to the English political theorist and sociologist Leonard Trelawney Hobhouse (1864–1929). He was considered one of the founders of sociology as a discipline and a proponent of social liberalism in the UK. The lecture series was first published in 1930 by Oxford University Press and subsequently replaced by the British journal of *Sociology*. The subject matters of the annual lectures given in his memory varied from ethics, biology, and anthropology to political theory, law and to religion, matching thus the diversity of Hobhouse's work. Stebbing—who had a full-time position at the University of London at Bedford College from 1920 on before being appointed professor there in 1933—knew Hobhouse personally, as she emphasized at the beginning of her lecture. Hobhouse, who had received his academic education in Oxford, held the newly established Chair of Sociology at the London School of Economics and Political Science (LSE), from 1920 on, where he remained until his death in 1929. According to Siobhan Chapman (2013, 166), Stebbing gave her Hobhouse lecture on 12 May 1943 in Cambridge, where the lectures of Bedford College and LSE were temporarily taking place.

Stebbing wrote with great respect about Hobhouse as someone whose philosophy had "informed" his life: the atheist (from a religious family) dealt with ethical and epistemic contexts as well as political-sociological topics. Similarly, for Stebbing, moral philosophy or ethics was not linked to religious faith, but to humanism and a materialist view of life, as is also evident from her text "Ethics and Materialism" (Stebbing 1939). In relation to political and social issues, Chapman (2013, 182) describes Stebbing as undoubtedly "left wing", which corresponds to her approach of critical, purposeful thinking and constant questioning of the process of the formation of opinion and the conscious or unconscious bias that characterizes public, scientific, or political debates. This was already an important element in Stebbing's *Thinking to Some Purpose* (Stebbing 1939) as a republican understanding of life in modern democracies—in the sense of Hannah Arendt—as Eric Schliesser has noted, also with reference to Stebbing's remarks in *Ideals and Illusions* (Stebbing 1941b; Schliesser 2023, n.d./ms.).

In *Ideals and Illusions*, one of the last books published in her lifetime, Stebbing claims that the democratic ideal is closely linked to the ethical principle that all

[14] Neurath's library is an enormous and significant source of his intellectual development, and documents in a detailed manner his engagement with his contemporary colleagues, friends, allies, and enemies as well. See further Stadler (2019) about the importance of Neurath's library.

people should be equally free and happy—and as Schliesser concludes, for this she deserves to be "re-inscribed in the genealogy of modern republicanism" (Schliesser 2023). Furthermore, Stebbing emphasized the importance of utopian ideals (or ideas) for social change, such as those represented by the internationalism of the League of Nations, in contrast to the influence of two other frequently cited factors, economic structure and the possession of power. According to Stebbing, it is a mistake to believe that the "achievements of science" can deliver these ideals and "lead us to paradise". To think in scientific "abstractions" when it comes to moral philosophy is to fail as a philosopher (Waddington 1941b, 8f.), a thought Stebbing expanded on in her lucid Hobhouse Lecture.

11.3.2 The Booklet

In her text, Stebbing contrasted moral or ethical statements with scientific statements and examined them as "isolates", such as those used in economics or as "abstractions", as used in the sciences. For her, the comparison of moral statements with scientific statements makes clear that they follow different rules and belong to different areas of human interaction: while moral judgements, which are concerned with "ethical situations", with "good" and "evil", are always individual and circumstance- or context-dependent, scientific judgements about the physical world can be objectified, tested and abstracted. In this sense, moral philosophy is the "least abstract" of all "systematic studies"; it has to be practically oriented, cannot be restricted to a linguistic analysis, and provide guidance for action. As she wrote at the beginning of *Ideals and Illusions*:

> Moral philosophers, I contend, must be concerned with the ways in which men live – their ways of life which embody their ideals. I conceive that it falls within the proper province of moral philosophers to formulate ideals worth living for and to attempt to make clear principles which may afford guides for action. This is a task singularly difficult to fulfil. (Waddington 1941b, 4)

According to Stebbing, statements on moral issues and analyses of such claims could only be made in the context of ideals and ideas as a whole, which express a certain way of life, and are connected to our practice. In this sense, she defended supporters of utopianism and idealism, for whom the practices and means of achieving an ideal must correspond to the moral standards that also determine their ends—otherwise they would be interchangeable with those of dictators like Hitler or Mussolini (Waddington 1941a, 19). Here she criticized in particular E. H. Carr in his book, *The Twenty Years' Crisis, 1919–1939* (Stebbing 1939), according to which only realism is an appropriate "scientifically" founded analysis of International Relations and appropriate response to political crises. At the same time, Stebbing emphasized that such ideals—although they are crucial and guiding our actions—are not invariable and do not apply equally everywhere at the same time and therefore require special attention and defense, as in the case of the struggle of democracies against fascism and national socialism (Waddington 1941a, 216f).

Stebbing's central claims in *Men and Moral Principles* followed at least in part G. E. Moore's view of ethics, as Chapman points out and as Stebbing herself confirmed in the text. But there are also clear differences as, for example, Siobhan Chapman has shown:

> It is difficult to see, for instance, how Moore's pronouncement that love and aesthetic appreciation can be known to be good and valuable for their own sake would stand up to Stebbing's complaint that moral philosophers 'have been mainly concerned to provide a list of goods, or a set of principles – or, worse still, a single good or a single principle – and have held this up to our admiring contemplation without asking what it would be like if everyone pursued this good [...]. (Chapman 2013, 167f).

And in fact, Stebbing's aim in *Men and Moral Principles* was seemingly a reform or revision of moral philosophy—distancing herself from the tradition of Kant or the hedonists, or those who have seen the solution in abstraction (such as self-realization). Instead, her approach could be summarized with the following points:

1. *Subjects*: Moral philosophy concerns ethical situations (definition).
2. *Practicality:* Moral philosophy has to be practical and cannot stop at the pure linguistic level of analysis to provide guidance for action.
3. *Empiricism*: Moral philosophy must take into account the ways in which we live and the problems the world is confronting us with.
4. *Holism*: Actions are to be seen under broad descriptions and in connection with other things (overall situations, as "moral values depend upon non-moral values").

As already mentioned, moral issues cannot be dealt with in isolation, because 1–4. Morality must always take into account the circumstances and background assumptions, and rules cannot be applied blindly and universally. *Speak the truth* is good, but one can imagine circumstances in which one must act differently to act morally. One way to avoid the problem would be to list possible circumstances that prevent one from unwanted consequences and would thus liberate one from various constraints: *Speak the truth provided that you not thereby break a promise*; *wound your friend, betray your comrades, destroy someone's illusion,* or as Margaret Macdonald (1945) continues Stebbing's list *that you give dangerous information to a madman, destroy another's peace of mind unnecessarily.* These circumstances mitigate, or even diminish the seriousness and restrictive character of your moral rule and enable you to act differently. But after explicating all (or just most) of these extenuating circumstances, the moral rule loses its character of being a rule in the strict sense.

These often-unexpected conditions are dictated by life, and therefore one's own morals and inner rules, "ideals" and "ways of life" have to be called upon. By connecting moral actions to ways of life and ideals, Stebbing could be read as hinting at Aristotle's idea that one's moral character does not surface in particular actions, but in habits. Recurrence of certain forms of actions (that document and express ideals in the long run) are the hallmarks of moral characters and decisions, and Stebbing admitted that she found "most help" from Aristotle (and G. E. Moore).

11.4 Influences and Reception

"The Inadequacies of the Scientific Attitude" and *Men and Moral Principles* could be read in conjunction to offer an ethical or moral reframing of the scientific attitude, without absolutist, technocratic or hyperbolic connotations. Neurath certainly hinted at this idea when he wrote in a text published after his death in 1946:

> I clearly realized the tendency within our movement to deal with the actual life when I looked at Stebbing's "Ideas and Illusions", the preface dated Tintagel April 1941, where the school she and her friends had organized had been evacuated from London. Here she continued her fight against muddled arguing, as started in "Philosophy and the Physicists" and in "Thinking to some Purpose". But during the war also appeared her "A Modern Elementary Logic" which was intended to be a book for students, some of them in the army, without any guidance from a teacher. I speak of these details, because they clearly show, how persistent scientific life is. (Neurath 1946, 79)[15]

This passage and many other statements by Neurath show the substantive and personal connection between the two that developed from the 1930s onwards. They agreed that the role of philosophy, science and expertise must be reconsidered in the face of crises and that the formation of modern democracies, and an egalitarian world society must be actively promoted through humanistic ideals and a "scientific attitude"—as an independent, active attitude dealing with institutions, politics, or the media (see Silke Körber's chapter in this volume).

In the 1930s and 40s, Stebbing published in the most important philosophical journals in the UK as well as various specialist and non-fiction books, was a member of relevant societies and groups and was also known to a wider public—as a woman she had a special position in British philosophy at the time. Despite her intellectual work, Stebbing was not accepted into the canon of mainstream philosophers after her early death, and her influence remained local and restricted in several ways. Because of her unique position in her time, Stebbing has not yet been seen as an actor in feminist philosophy, although she certainly paved the way for others.[16] Her ethical and socio-political writings and activities have long been neglected in favor of logic but have recently received more attention.

Besides the works mentioned in the first footnote, two important monographs have appeared recently about female British philosophers from the late 1940s on, for example on their reform of moral philosophy. Those books explain and debate

---

[15] This passage is indeed important and telling about Neurath's and Stebbing's relationship. It is quoted also in Chapman (2013, 165).

[16] Very unfortunately, philosophers could not always overcome the fact that Stebbing was a woman. In a quite negative review, C. D. Broad wrote about the mistakes of Stebbing's views on Jeans and Eddington's ideas. But even when he applauded the general critical task ("cleaning up the mess") that Stebbing overtook to handle the physicists' philosophy, he compared that critical-conceptual "intellectual scavenging" (that was obviously much required from someone, although no one dared to spare their important time for that) to house-cleaning and wrote that "but at the end of it, [Stebbing] must have enjoyed something of the exhilaration of a good housewife who has at last completed her spring-cleaning" (1938, 226).

how narratives are created, and internal and external factors influenced the field at the time. One of them, *The Women Are Up to Something*, by Benjamin Lipscomb (2022, 74), mentions Stebbing only as an isolated figure similar to Dorothy Emmet, Mary Glover, and Martha Kneale. The other volume, *Metaphysical Animals* by Clare Mac Cumhaill and Rachael Wiseman (2022) goes into more detail about Stebbing's interaction with members of the Vienna Circle and her approach to the subject of logic, analyses, and moral philosophy, and provides an interesting explanation for how her importance has been underestimated or forgotten overall, and especially in Ayer's *Language, Truth and Logic* that neglected Stebbing's approach and became a dominant influence:

> What remained, in Ayer's vision of philosophy, was a *technique*. Philosophers analyse propositions and scientists verify them. Gone was Stebbing's and Moore's vision of modern logic as a source of metaphysical insight. Gone too was Stebbing's conviction that a general public educated in linguistic analysis would be better able to resist propaganda, and better able to understand the world around them. Ayer repurposed the methods of analysis in a way that made the everyday world alien and inscrutable. (Mac Cumhaill and Wiseman 2022, 55).

As a result of the lack of an established network within the existing disciplinary structures and of a new philosophical direction, the historiography of this philosophical current was also impacted and biased, which was particularly bitter as it was written by an avowed friend of Stebbing's, Gilbert Ryle:

> Susan Stebbing, who had died in 1943, would soon be written out of the story, along with Margaret MacDonald (the post-war editor of Stebbing's Analysis) who passed away just as the first official history *The Revolution in Philosophy* (1956), edited by Gilbert Ryle, went to print. Neither of these women gets a mention – an inexcusable omission by Ryle, who knew Stebbing and MacDonald, and their work, well. The 'revolution in philosophy' had been won bloodlessly, by a combination of mortality and studied neglect. (Mac Cumhaill and Wiseman 2022, 135).[17]

However, the immediate reception of Stebbing's publications was sometimes not positive either. In the case of *Men and Moral Principles* one reviewer, the biblical-scholar R. H. Fuller (1944), basically just told the reader the content of the booklet in two sentences, and concludes that "Prof. Stebbing gives brief examples of how the moral philosopher ought rightly to proceed, but here she is meagre and disappointing."[18] As a priest, Fuller might have expected more concrete suggestions from Stebbing and found the text lacking in that respect. But it was not just him; Margaret Macdonald (1945, 77), a student of her, who became very influential after

[17] It has to be noted though that both J. O. Urmson (in his *Philosophical Analysis: Its Development between the Two World Wars*) and John Passmore (in his *A Hundred Years of Philosophy*) have discussed Stebbing's views in their influential historical accounts of philosophy. See Urmson (1956) and Passmore (1957).Both works focus on Stebbing's early logical work and not on her later, socially engaged, critical views. We are grateful to Michael Kremer for this suggestion.

[18] About *Ideals and Illusions*, the known British philosopher John Laird had written in the obituary volume of Stebbing that there was a "general agreement" in 1948, that the book "though nothing like a failure, is somewhat below the very high level of [Stebbing's] best work" (Laird 1948, 20).

Stebbing's death,[19] also thought that Stebbing left "unexplained" the most important points of her view about ethical decisions and their grounds.

As philosopher Ernest Nagel and one of her colleagues noted, Stebbing had to deal with many responsibilities and an overload—without the advantage of having the private or professional infrastructure that her male colleagues had at their disposal (Mac Cumhaill and Wiseman 2022).[20] This and her illness with cancer not only had an impact on the number of publications. In the case of *Men and Moral Principles* it affected perhaps also her ability to develop a positive programme that was more detailed. Stebbing's text remained largely unrecognized. In principle, it could have been more effective. A decade later G. E. M. Anscombe (1958)—in the focus of both books mentioned above—published a similar, more programmatic and most influential paper, "Modern Moral Philosophy" in the same journal where Macdonald's review has appeared. Anscombe was deeply critical of the whole line of British tradition of moral philosophy from Hume and Mill to Sidgwick and Moore and fostered an Aristotelian turn that consisted in leaving behind old remnants of moral philosophy that stem from the vocabulary of *moral* obligation, *moral* ought, *moral* duty and related terms—according to her, these should be replaced by an "adequate philosophy of psychology". Some of these ideas might be similar to Stebbing's thought, like the search for more psychological and empirical considerations and the emphasis on Aristotle, while others point to entirely different directions. Although Anscombe did not mention Stebbing in the paper, we might plausible assume that she would have belonged to the old tradition of British moral philosophy in Anscombe's eyes with her discussion on morally good and wrong.

As mentioned before, there might be numerous factors in the long negligence of Stebbing that started after her death—it would be somewhat misleading to deny the prominence of Stebbing during her life-time given her institutional roles, and countless appearances at conferences, symposiums, and the broad discussion of her work. In fact, her work in logic was much praised since the early 1930s and many held it in high esteem (Chapman 2013, 172–173), logicians and philosophers alike, even after World War Two. As Margaret Macdonald (1945, 76) noted, Stebbing was among those rare, almost unique group of philosophers whose "philosophy informed [their] life", and thus her logic shall be seen in connection to other works. Yet again, Neurath could not have agreed more.

---

[19] There is an increasing scholarship on MacDonald recently (with some hint to Stebbing as well), see Kremer (2022), Spinney (2023), Vlasits (2022), West (forthcoming), Kremer and Misak (forthcoming).

[20] Being a woman clearly closed many important ways from Stebbing. In 1938, Stebbing applied for Moore's chair in Cambridge; although it would have been a huge advancement for her academically and personally/existentially at the same time, she did not get it. As Ryle told her, "Of course everyone thinks you are the right person to succeed Moore, except that you are a woman" (quoted in Chapman 2013, 126). After all, Ludwig Wittgenstein was elected to the Cambridge Professorship. As Stebbing commented, after noting how "distinguished" Wittgenstein was, "only I do not think he is good for students and he has a bad effect on lecturers" (Stebbing to Neurath, 14 February 1939, ONN/304).

11.5 Editorial Issues

"The Inadequacies of the Scientific Attitude" is kept in Neurath's archive, Otto Neurath Nachlass, Wiener Kreis Archiv, Rijksarchief in Noord-Holland, Haarlem, The Netherlands, Folder 400/R.76. *Men and Moral Principles* was published originally by Oxford University Press—we are grateful for the Archive and the Press for permission to use both texts.

The copyright of the materials of Susan Stebbing belongs to her great nieces, Alison Willott and Celia Pyke-Lees, who kindly agreed to our republishing these texts. We are more than grateful to them for the generosity and kindness to support our research and help the community by letting these works of Stebbing go public.

In "The Inadequacies...", we provided a few editorial notes about the mentioned scholars and figures to help the reader in situating the text and its references, and always did so in footnotes within [[double square brackets]]. In that text, we also reproduced one marginal note of Stebbing in [single square brackets].

In *Men and Moral Principles* we did not introduce any editorial notes as it was published already after Stebbing's death, quite surely with her full approval. Stebbing's original footnotes are regular notes now as well, and we have inserted only two things, namely Neurath's marginal notes from his copy of the booklet, now within [[double square brackets]], and the information about the original publication within [single square brackets].

Acknowledgments We are grateful to Siobhan Chapman for helping us to decipher a few phrases from Stebbing's handwriting in "The Inadequacies..." text. Seemingly, as time went on during the preparation of the text, Stebbing's handwriting became worse and less readable. Without Prof. Chapman's experienced judgment, we would have not been able to finish our work. We are also indebted to Michael Kremer for his notes and comments, and to Friedrich Stadler and Georg Schiemer for providing space to make these texts available in this volume, and we are grateful for their patience and encouragement.

References

Anscombe, G. E. M. 1958. Modern Moral Philosophy. *Philosophy* 33 (124): 1–19.

Beaney, Michael. 2003. Susan Stebbing on Cambridge and Vienna Analysis. In *The Vienna Circle and Logical Empiricism*, ed. Friedrich Stadler, 339–350. Dordrecht: Springer.

———. 2016. Susan Stebbing and the Early Reception of Logical Empiricism in Britain. In *Influences on the Aufbau*, ed. Christian Damböck, 233–256. Dordrecht: Springer.

Beaney, Michael, and Siobhan Chapman. 2017/2021. Susan Stebbing. In *The Stanford Encyclopedia of Philosophy*, ed. Edward N. Zalta and Uri Nodelman. https://plato.stanford.edu/archives/fall2022/entries/stebbing/.

Broad, C. D. 1938. Review of L. S. Stebbing, *Philosophy and the Physicists*. *Philosophy* 13 (50): 221–226.

Burke, Christopher, and Wim Jansen. 2022. *Soft propaganda, special relationships and a new democracy: Adprint and Isotype 1942-1948*. Amsterdam: Uitgeverij de Buitenkant.

Burke, Christopher, and Adam Tamas Tuboly. 2024. *Otto Neurath in Britain*. Cambridge: Cambridge University Press.

Cat, Jordi, and Adam Tamas Tuboly, eds. 2019. *Neurath Reconsidered: New Sources and Perspectives*. Cham: Springer.

Chapman, Siobhan. 2013. *Susan Stebbing and the Language of Common Sense*. Basingstoke: Palgrave Macmillan.

Coliva, Annalisa. 2021. Stebbing, Moore (and Wittgenstein) on Common Sense and Metaphysical Analysis. *British Journal for the History of Philosophy* 29:914–934.

Collins, Harry. 2023. The Most Important Thing about Science is Values. *Interdisciplinary Science Reviews* 48 (2): 264–275.

Collins, Harry, and Robert Evans. 2017. *Why Democracies Need Science*. Cambridge: Polity Press.

da Cunha, Ivan Ferreira. 2024. Otto Neurath's Modernist Utopianism: Linking the Vienna Circle and H.G. Wells. *HOPOS: The Journal of the International Society for the History of Philosophy of Science* 14 (1): 25–51.

Douglas, Alexander X., and Jonathan Nassim. 2021. Susan Stebbing's Logical Interventionism. *History and Philosophy of Logic* 42:101–117.

Franco, Paul L. 2023. Susan Stebbing on Logical Positivism and Communication. *Ergo: An Open Access Journal of Philosophy* 10 (48): 1378–1402.

Fuller, R. H. 1944. Review of Stebbing's *Men and Moral Principles*. *Theology* 47 (290): 191.

Janssen-Lauret, Frederique. 2017. Susan Stebbing, Incomplete Symbols, and Foundherentist Meta-Ontology. *Journal for the History of Analytical Philosophy* 5:6–17.

———. 2022. *Susan Stebbing*. Cambridge: Cambridge University Press.

———. 2024. Grandmothers and Founding Mothers of Analytic Philosophy: Constance Jones, Bertrand Russell, and Susan Stebbing on Complete and Incomplete Symbols. In *Russell and Women*, ed. Landon D. C. Elkind and Alexander Klein, 207–239. London: Palgrave.

Körber, Silke. 2019. Thinking about the 'Common Reader': Otto Neurath, L. Susan Stebbing and the (Modern) Picture-Text Style. *In Cat and Tuboly* 2019:451–470.

———. 2023. *Die Visualisierung von Wissen im "Jahrhundert des Auges": Otto Neurath, Isotype und Adprint*. Cham: Springer.

Körber, Silke, and Adam Tamas Tuboly. 2025a. Thinking Clearly for a Modern Democracy: Susan Stebbing and Otto Neurath. In *Susan Stebbing on Logic and Analysis*, ed. Siobhan Chapman and Brian Garvey. Basingstoke: Palgrave Macmillan. forthcoming.

———. 2025b. *The Selected Correspondence of Susan Stebbing and Otto Neurath*. Unpublished manuscript.

Kremer, Michael. 2022. Margaret MacDonald and Gilbert Ryle: A philosophical friendship. *British Journal for the History of Philosophy* 30 (2): 288–311.

Kremer, Michael, and Cheryl Misak. forthcoming. *"So first hand": Analytic Philosophy in the Letters and Writings of Margaret Macdonald, 1932–1937*. Unpublished manuscript.

Laird, John. 1948. Reflections Occasioned by *Ideals and Illusions*. In *Philosophical Studies: Essays in Memory of L. Susan Stebbing*, 20–30. London: George Allen & Unwin.

Lipscomb, Benjamin. 2022. *The Women Are Up to Something: How Elizabeth Anscombe, Philippa Foot, Mary Midgley, and Iris Murdoch Revolutionized Ethics*. Oxford: Oxford University Press.

Mac Cumhaill, Clare, and Rachael Wiseman. 2022. *Metaphysical Animals: How Four Women Brought Philosophy Back to Life*. New York: Doubleday.

Macdonald, Margaret. 1943. L. Susan Stebbing. *Girton Review* 123 (Michaelmas Term, 1943): 14–17.

———. 1945. Review of Stebbing's *Men and Moral Principles*. *Philosophy* 20 (75): 76–78.

Moravec, Matyáš, and Peter West. 2023. Stebbing and Eddington in the Shadow of Bergson. *History of Philosophy Quarterly* 40 (1): 59–84.

Neurath, Otto. 1944. *Foundations of the Social Sciences*. Chicago: The University of Chicago Press.

———. 1946. After six years. *Synthese* 5:77–82.

Passmore, John. 1957. *A Hundred Years of Philosophy*. London: Duckworth.

Pickel, Bryan. 2022. Susan Stebbing's Intellectualism. *Journal for the History of Analytical Philosophy* 10:1–24.

Schliesser, Eric. 2023. Illusions, and the forgotten history of analytic political philosophy. *Digressions&Impressions* blogspot, June 14, 2023. https://digressionsnimpressions.typepad.com/digressionsimpressions/2023/06/ideals-and-illusions-susan-stebbings-1941-moving-wartime-work-published-while-she-and-the-kingsley-lodge-school-for-girls.html

———. n.d./m.s. *Stebbing on Clarity*. unpublished manuscript.

Spinney, Oliver Thomas. 2023. Margaret Macdonald on the Argument from Dreaming. *The Philosophical Quarterly* 74 (3): 962–977.

Stadler, Friedrich. 2019. A Viennese Library in Exile: Otto Neurath and the Heritage of Central European Culture in the Anglo-Saxon World. *In Cat and Tuboly* 2019:23–44.

Stebbing, L. Susan. 1934. *Logic in Practice*. London: Methuen & Co.

———. (1937). *Philosophy and the Physicists*. London: Penguin Books.

———. 1939. Ethics and Materialism. *Ethics* 50 (1): 35–44.

———. 1941a. The Relations between Science and Ethics. *Nature* 148:277–278.

———. 1941b. *Ideals and Illusions*. London: Watts and Co.

Tuboly, Adam Tamas. 2020. Knowledge Missemination: L. Susan Stebbing, C.E.M. Joad, and Philipp Frank on the Philosophy of the Physicists. *Perspectives on Science* 28 (1): 1–34.

Urmson, J. O. 1956. *Philosophical Analysis: Its Development between the Two World Wars*. Oxford: Oxford University Press.

Vlasits, Justin. 2022. Margaret MacDonald's scientific common-sense philosophy. *British Journal for the History of Philosophy* 30 (2): 267–287.

Waddington, C. H. 1941a. *The Scientific Attitude*. Harmondsworth: Penguin Books.

———. 1941b. The Relations Between Science and Ethics. *Nature* 148:270–274.

———., ed. 1942. *Science and Ethics*. London: Unwin Brothers Limited.

Waddington, C.H. 1948. *The Scientific Attitude*. Second and revised edition. Harmondsworth: Penguin Books.

Waddington, C.H., A.C. Ewing, and C.D. Broad. 1941–1942. Symposium on the Relations between Science and Ethics. *Proceedings of the Aristotelian Society* 42: 65–100, 100A–100H.

Werskey, Gary. 1978. *The Visible College: A Collective Biography of British Scientists of the 1930s*. London: Allen Lane.

West, Peter. 2022a. L. Susan Stebbing Philosophy and the Physicists (1937): A re-appraisal. *British Journal for the History of Philosophy* 30 (5): 859–873.

———. 2022b. The Philosopher Versus the Physicist: Susan Stebbing on Eddington and the Passage of Time. *British Journal for the History of Philosophy* 30 (1): 130–151.

———. forthcoming. Philosophy is Not a Science: Margaret MacDonald on the Nature of Philosophical Theories. *HOPOS: The Journal of the International Society for the History of Philosophy of Science*. Forthcoming.

Chapter 12
The Inadequacies of the Scientific Attitude

L. Susan Stebbing

'Science by itself and so far as I can see only science by itself, unadulterated by mixture with any contrary ideal, is able to provide mankind with a way of life which is, firstly, self-consistent and harmonious, and, secondly, free for the exercise of that objective reason on which our material progress depends' (p. 124). This claim is made by Dr. Waddington in his interesting but in some ways absurd little book: *The Scientific Attitude.*[1]

I shall first summarise what I take to be his main contentious: then I shall venture to make some criticisms: finally, I shall try to show that the universal adoption of 'the scientific attitude' would not suffice to cure the ills of our world.

Dr. Waddington's starting point is naturally provided by what we have become accustomed to call 'the breakdown of civilization'. I assume that it is common ground to us all that there is a good deal very wrong with our society, and further—which is what is really important—that much of that wrong is remediable. Dr. Waddington, correctly, in my opinion, does not *argue* that this is so; he takes it for granted, and then proceeds to propound his remedy. This remedy is—adopt the scientific attitude and let science take control.

The adoption of this remedy is not so simple as Dr. Waddington naively supposes; nor, indeed, is his statement of the remedy so clear as to give us much confidence in his scientific guidance. Lest you should think that I exaggerate I will quote one paragraph in full:

L. Susan Stebbing died before publication of this work was completed.

[1] [[Page numbers refer to the first edition as C.H. Waddington, *The Scientific Attitude*, London: Penguin Books Ltd., 1941a. The manuscript is reproduced by the kind permission of the copyright holders of Stebbing's materials, Alison Willott and Celia Pyke-Lees.]]

L. S. Stebbing
1885–1943
London, UK

G. Schiemer (ed.), *The Legacy of the Vienna Circle*, Vienna Circle Institute Yearbook 30, https://doi.org/10.1007/978-3-031-80568-4_12

> There is no other intellectual movement [than science] at the present time which has been successful enough to have the confidence necessary to carry through the struggle against the barbarism of the Nazis. We may defeat them physically by sheer dogged pride, but unless we have something to put in place of their ideas, we shall win the war only to find that we have gone fascist ourselves. Religion? – it has produced a few Niemollers, and will continue to do so, but its hold over society is too weak now to beat down this attack on all that it values. Capitalism? – who believes in pure capitalism except a few old men in clubs, and who knows just what amount of doctoring would be required to make it work? Democracy – yes, but what is democracy to-day? something which has brought us to our present state; not good enough as it stands. Socialism? again yes, but whose brand? 'we are all socialists now'. Communism – it is the only living system, other than science, which is nearly consistent enough to do the job, but like Christianity, too few people believe in it, and, near though it is to science, I shall give some reasons later why I think that science is better; but it is the only serious competitor in sight. Science, I believe, could fill the gap. (p. 65)

The gap to be filled is—you will have noticed—the gap left—when we have defeated the Nazis—by the defeat of 'their ideas'. Five competitors for filling the gap are mentioned: Religion, Capitalism, Democracy, Socialism, Communism—each to be in turn rejected although fairly full marks are awarded to Communism, which is clearly regarded by Dr. Waddington as *proxime accessit*. Science alone is left. What exactly, then, does Dr. Waddington mean by "Science", and what is it that Science can *do* that the other five 'competitors' cannot do?

I assume that by "science" Dr. Waddington means "the natural sciences", and in particular the biological sciences—in which, I believe, he includes sociology. What, then, is it that natural science is to *do*, to fill the gap. The answer appears to be given in the statement which I quoted at the beginning of my paper: science can provide mankind with a way of life—in fact, with what it is not uncommon to call an *ideal*. Science, as it is further maintained—science *alone* can provide us with an ideal that is 'self-consistent and harmonious,' and 'free for the exercise of that objective reason on which our material progress depends'. It is important to notice exactly what the claim is: science *provides* the ideal; it is not merely concerned with showing us what steps must be taken to realise an ideal not completely given by science. Dr. Waddington is fully aware that in making this claim he is talking about ethics; indeed, he is anxious to insist that science tells us what good and evil are.

Before we examine what science has to tell us on this topic we must clear away an objection which, if it could be sustained, would prevent Dr. Waddington from getting under weigh at all; consequently, he seeks at the outset to dispose of the objection—namely that science is ethically neutral.

Chapter II is entitled "Science is not neutral". To those of you who are interested in the claims made by Dr. Waddington, I commend this chapter as deserving of careful study. The extraordinary confusions which abound in it need to be very carefully unravelled. To do this properly would take longer than the time at my disposal. I must, therefore, condense. It will, I think, be clearer if I begin by saying what I understand by the statement: "Science is ethically neutral". Bertrand Russell, I believe, first introduced this phrase (in his Herbert Spencer lecture on 'The Scientific Method in Philosophy'). He left us in no doubt what he meant, namely, that a scientific theory is not dependent for its truth, or falsity, upon our aspirations. For instance, to take an example of my own, the truth of the second law of thermodynamics—if it be true—has nothing to do with the fact that some people cannot bear to believe it, nor with the fact that Dean Inge[2] approves of it. What we *want*, what we *desire*, what we judge *ought* to be—these have no relevance to natural science. Dr. Waddington does, I think, admit this, for he says: 'Science as a whole certainly cannot allow its judgement[s] [about facts] to be distrusted by ideas of what ought to be true, or what one may hope to be true' (p. 25). (I wonder why he has to qualify this remark by 'science *as a whole*'?) But this statement immediately follows his denial that science is ethically neutral, the denial takes the form of asserting that scientists do 'more or less passionately hope to be able to prove or disprove some particular theory on which these are working'. The point of this contention is that Dr. Waddington wants to insist that 'the scientific attitude is as full of passion, as much a function of the whole man and not merely of an intellectual part of him, as any other approach to human action.' The last seven words—'as any other approach to human action'—are revealing; they show that he is thinking of those sciences that are concerned with human behavior, biological sciences and social sciences. Still, I hope he would be equally anxious to insist that a mathematical physicist 'more or less passionately' hopes to be able to prove the theory upon which he is working. If, for instance, you have read Leopold Infeld's most interesting autobiography, *Quest*, you will remember how well he conveys the physicist's absorption in his work, and the excitement with which he pursues his task. Infeld's account of Einstein at work makes this vividly clear to us. But this passionate excitement of the scien*tist* has nothing whatever to do with the ethical neutrality of *science*.

You may have noticed that I have just spoken of the *ethical neutrality of science* and of the passionate excitement of the *scientist*. This is a distinction upon which I shall want to lay a great deal of stress later on in my paper; it is a distinction almost entirely ignored by Dr. Waddington—the distinction between *science* and *scientists*.

To return to Dr. Waddington's characterisation of the scientific attitude. This, he says, 'the matter of fact as against the romantic, the objective as against the subjective, the empirical, the unprejudiced, the *ad hoc* as against the *a priori*' (p. 47). How naive this is, how odd, how unduly simplified! In allowing myself these three exclamatory comments, am I being unscientific? I leave you to judge.

What, then, does Dr. Waddington's characterisation of the scientific attitude amount to, and in what sense does he maintain that science is not ethically neutral? The scientific attitude is matter-of-fact, objective, empirical, and unprejudiced; if I allow myself to accept a theory because it would confirm my hopes, or to reject a theory because it runs counter to my desires, then I am failing to maintain the scientific attitude. I hope no one will want to deny this. The scientific *attitude* is dispassionate; ethical considerations are simply irrelevant to it. The scien*tist*, however, is

[2] [[William Ralph Inge (1860–1954) was an English author, professor in Cambridge, president of the Aristotelian Society (1920–21), and dean of St Paul's Cathedral in London. Inge was a prolific author with more than three dozen books, scholarly papers, opinion pieces, and he was nominated for the Nobel Prize in Literature three times. He also wrote about science, truth, and natural science.]]

a human being, a 'whole man' so to speak, a *person* who wants to find out *how* things work and who also takes delight in the *process* of finding out, so that he feels excited when his theory is 'coming out all right'. Again I hope that no one will want to deny this. But how does the fact that a scientist, who is a human being, takes delight both in his work and in his personal success affect the question whether science is ethically neutral? I do not think it affects it at all. Nor does Dr. Waddington in spite of the fact that he introduces this consideration into the centre of his discussion of this topic.

So far as I can discover, Dr. Waddington wants to make two points: (1) scientific discoveries are extremely relevant to the whole situation upon which we may have to pass ethical judgements; (2) science itself implies an ethical standard.

These two points must be separately discussed, for they merit the most serious consideration.

(1) Take as an example scientific discoveries concerning food values. He has no difficulty in showing that the amount and kind of nutrition contained in a given food, (i.e. its food-*value*) can be determined by scientists, and that the facts thus ascertained are relevant to the decision whether we want to produce that food provided we want to secure that amount and kind of food value. But these scientific facts, i.e. the knowledge thus provided by scientific investigation, do not in the least determine whether we ought to produce that particular food, nor, having produced it, to whom the food should be distributed, and in what manner.

Consider another, and for our purpose, an even clearer example of the relevance of scientific discoveries to our ethical judgements. It is now known that the most prevalent cause of bone tuberculosis in children is due to milk containing tubercle bacilli. It is further known what steps must be taken to prevent the selling of infected milk. We could, if we would, in a very short time free this country from children's tuberculosis. No doubt one reason why the obvious and effective steps to secure this end have not been taken is that there is no public demand for them to be taken. This failure on the part of the public is largely due to ignorance of the facts now well established by scientists.

It is with reference to such facts as these that Dr. Waddington writes as follows: 'Our ethical notions are fundamentally based on a system of individual responsibility for individual acts. The principle of statistical correlation between two sets of events, although accepted in scientific practice, is not usually felt to be ethically completely valid. If a man hits a baby on a [the] head with a hammer, we prosecute him for cruelty or murder: but if he sells dirty milk and the infant sickness or death rate goes up, we merely fine him for contravening the health laws' (26). These contentions are true. What do they show? In my opinion they show that our ethical judgments may be faulty through lack of information with regard to factual situations, the information being of the kind that can be provided for us by the research of scientists, and in no other way. Thus in the example given by Dr. Waddington, scientists have shown that the death of babies from tuberculosis is the causal convergence of tubercle bacilli absorbed from milk; they have further shown how this danger can be wholly removed. In short, the research of certain scientists has provided us with information with regard to the remote, and the near, responsibility of certain sets of people for the death and disease of a number of children. *If* we judge that premature death and present suffering are evil, and ought not to be tolerated, then this information is of the utmost importance in enabling us to pass moral judgments about definite states of affairs. Nothing of this, however, has the least bearing upon the question whether or not scientific research is ethically neutral. The scientist, as Dr. Waddington elsewhere maintains, must accept the facts without letting them be 'distorted by ideas of what ought to be true, or what one may hope to be true'.

It is of great importance to make this point clear. The decision with regard to what *ought* to be done—in the case of permitting the unrestricted sole of tubercle infected milk—is no more specially within the competence of the scientist than of, say, the poets, or the ordinary man who professes no 'specialism'. If a given scientist has views on this matter—as Dr. Waddington has—it is not in his capacity as a scientist but as a humanitarian citizen. To speak more generally, if a scientist has views with regard to the frustration of science, and feels a special responsibility for providing knowledge that could be put to nefarious uses, it is *again* not *as a scientist* but *as a citizen* or *as a good man* that he holds these views. A scientist anxious to prosecute his research may be quite indifferent to the suffering its success entails upon people. (I believe this point was made by the American novelist, Sinclair Lewis,[3] in his novel 'Martin Arrowsmith', the main plot of which turned upon the temptation experienced by Martin Arrowsmith, a medical scientist, to experiment on people suffering from a virulent disease. I am afraid, however, that I do not remember the novel well enough to be able to cite it clearly. But now doubt you will be able to think of plenty of examples yourselves.)

Take as an example the discovery by the German chemist, Haber,[4] that nitrogen and hydrogen when heated together at high pressures in the presence of a catalyst such as iron or aluminium oxides form ammonia. This is an example of a purely scientific discovery. I do not doubt that Haber felt as much excitement in making

[3] [[Sinclair Lewis (1885–1951) was an US fiction writer and journalist, who travelled through Europe, visited Germany during the rise of the Nazis, worked as a journalist and translator in New York and as Jack London's private secretary. In 1930, he was the first American to be awarded the Nobel Prize for Literature for his socially critical novels such as *Babbit*, *Arrowsmith*, and *Elmer Gantry*. His works enjoyed great popularity at the time and were made into movies, but were later almost forgotten. The dystopian novel *It Can't Happen Here* (1935), received greater interest again in view of the rise of Donald Trump and the polarisation of American society.]]

[4] [[Fritz Haber (1868–1934) was a German chemist and Nobel Prize winner for chemistry in 1918. As founding director, he headed the Kaiser Wilhelm Institute for Physical Chemistry and Electrochemistry in Berlin for over 20 years. His work on the catalytic synthesis of ammonia from the elements nitrogen and hydrogen laid the foundation for the Haber-Bosch process. This enabled the mass production of nitrogen fertiliser and thus the nutrition of a large part of the world's population. However, Haber's experiments with phosgene and chlorine gas shortly after the start of the First World War made it possible to use poison gas as a weapon of war. After the Nazis seized power, he emigrated to England in 1933, but died shortly afterwards.]]

this discovery as Kékulé[5] felt in working out the foundations of his Benzene theory—which formed the basis of his theory of the constitution of aromatic compounds. Now, Haber's discovery enabled the Germans to make their agriculture independent of Chile saltpetre during the war of 1914–18. This discovery no doubt prolonged the war, and on this we may well pass ethical judgments, but there is nothing in the least ethical, or moral, in the scientific discovery itself. The discovery is *amoral*, or if you prefer, non-moral.

Consider, as a trial example, one which I take from a recent article in *Nature*, by Dr. Waddington, on 'The Relations between Science and Ethics'.[6] He cites as a 'typical ethical judgement' the judgement: 'You are an animal of such a kind that you must consume 7 mgm of vitamin C per diem, and should consume 100 mgm'.—This, he says, is an ethical statement and also 'a statement which has scientific significance'. Now *what* in it is scientific and *what* is ethical? What is scientific (not the quite precise sense that scientists have made the discoveries noted in the sentence) is the statement that 7 mgm of vitamin C daily is necessary to keep such an organism as a human being alive and 100 mgm is necessary to keep the organism in good health. Dr. Waddington's use of the two words "must" and "should" in his sentence is quite in accord with English usage. But the words are ambiguous, for, if they be used merely to assert correlations scientifically established, they can be replaced by some such phrases as "is necessary" or "is in fact needed" without changing the scientific significance of the statement. The *ethical* valuation comes in only because Dr. Waddington is *also* using the words "must" and "should" to say what *ought* to be done. Being an animal of such a kind that 7 mgm of vitamin C daily is necessary to keep you alive, you must (i.e. *ought*) to consume that amount; and indeed, you *ought* to consume 100 mgm'. If I ask—"*Why* ought I?" the answer surely is: 'Because otherwise you will die, and will not keep in good health'. And if I go on to ask—'and why *ought* I keep in good health?'—the answer can no longer be given in terms of the human organism's need for vitamin C. Indeed, in these days, it is not difficult to imagine a situation in which the mother of a family of small children might judge that she *ought* to forego her daily quota of vitamin C in order that her children may have sufficient for their bodily needs. If she did so judge, her judgement would be *based upon* her acceptance of the scientific statement about vitamin C, and would not be in the least unscientific or anti scientific. It is not, however, a scientific judgement at all, but an ethical one.

I conclude this point by remarking that many of our ethical judgements are based upon judgements belonging to biological chemistry, and other sciences. In saying

---

[5] [[Friedrich August Kékulé (1829–1896) was a German chemist who laid the foundations for modern structural formulas in organic chemistry. His work played a decisive role in the development of organic chemistry and resulted in a boom in the German chemical industry, especially in the manufacture of artificial pigments. In 1858 he became Professor of Chemistry at the University of Ghent in Belgium and in 1867 accepted an appointment at the University of Bonn, where he lived until his death in 1896.]]

[6] [[See C.H. Waddington (1941b), "The Relations Between Science and Ethics", *Nature* 148 (September 6, 1941): 270–274.]]

that they are *based upon* these scientific judgements I am saying that the ethical judgement could not be made unless we had the information supplied by the scientists. In saying that, nevertheless, these scientific judgements are not *ethical* judgements I am saying that the ethical judgements do not constitute part of the subject-matter of the science in question. We may easily fall into the mistake of supposing the contrary, because scientists are not *merely* scientists; they are *also* human beings. Their specialist studies provide them with special information highly relevant to the determination of what, in a given situation, we ought to do provided that we desire such and such a state of affairs. That is to say, scientists are often able to give us help in knowing how to achieve our ends; but the scientist—in his capacity of a scientist—does not determine these ends.

(2) This brings me to the second point, which I said Dr. Waddington is concerned to make. This is that 'science itself imply an ethical standard'. In the chapter designed to show that 'science is not neutral' Dr. Waddington says 'those things which are based on the scientific attitude are good, those which stultify or deny it are to that extent bad'.[7]

I agree that to possess the scientific attitude is good; that to seek to stultify or deny it is bad. But I do not agree that these judgements of value are themselves judgments belonging to biology, or any other science. Dr. Waddington possibly fails to realise this because he will talk about science when he should be talking about *scientists* provided that he wants to make his views clear to us. A scientist will naturally believe that the scientific attitude is good, for the study of science is good. Further, scientific discoveries can be made only if the investigator is intelligent, capable of impartial and dispassionate thinking, submissive to fact, and—usually—able to cooperate with other scientists. These characteristics are, I should agree, valuable characteristics. Would that we all had them! They are by no means to be found only among scientists; historians also must possess them; a classical scholar, a specialist in sanskrit—indeed any scholarly student must have these characteristics to the extent to which he is able to carry on his work. They are characteristics so good for human beings to possess that any way of life that tends to their deliberate extinction is so far evil. We shall all, I hope, agree with the following statement of Dr. Waddington's:

> Under the impact of recent events there is a very remarkable agreement among scientists throughout the world that a system of thought such as Nazism is incompatible with the scientific temper and is, for that reason among others, to be ethically condemned (p. 26).

I dissent only from the comment that the agreement is 'remarkable'. But this consideration that the suppression of the scientific attitude is evil, is not a judgement of *science*; it is, I repeat, an ethical valuation of the characteristics needed by a scientist in order that he may carry on the work of science. When Dr. Waddington says: 'The maintenance of a scientific attitude does in fact imply the assertion of a certain ethical standard', he is saying something to which, at first sight, I might be willing to

[7] [Stebbing's note on the margin: "Perhaps I should quote the whole paragraph (p. 29)?"]

agree. For the statement is not very clear. If it meant that maintaining a scientific attitude implies possessing certain valuable characteristics, then I agree, and have already stated my agreement. But if it be meant that the maintaining of a scientific attitude *entails* an ethical standard, then that seems to me to be absurd. And I think this is what Dr. Waddington does mean.

This position is made, perhaps, a little clearer in his attitude in *Nature*, dealing—you will remember—specifically with the relation of ethics and science. He there says: 'we must accept the direction of evolution as good simply because it *is* good according to any realist definition of that concept' (op cit. p. 273). I do not profess to know what he means by the world "realist" here, but I think it is clear that he means that "good" is synonymous with "in the direction of evolution". 'In the world as a whole', he says, 'the real good cannot be other than that which has been effective, namely that which is exemplified in the cause of evolution'. He adds: 'It is, then finally clear that science is in a position to make a contribution to ethics, since ethics is based on facts of the kind with which science deals. And the nature of science's contribution is also clear; it is the revelation of the nature of the character and direction of the evolutionary process in the world as a whole, and the elucidation of the consequences, in relation to that direction, of various courses of human action' (273).

To me it seems clear that this agreement is faulty. Until I read Dr. Waddington's writings I had supposed that the fallacies committed by Herbert Spencer had been so often so conclusively refuted that even the most aged of Victorians would not have revived them. I am sorry to find I am mistaken, and that the dead horses of the Victorian era have been resuscitated by a young Cambridge scientist. I suppose that work has to be done all over again. I have time here only to make a few brief comments.

The direction of the evolutionary process can hardly be said to the unilateral; further, if the second law of thermodynamics is fine, is to be interpreted as it usually is, then we must expect that the evolutionary movement will be followed in time by a devolutionary movement, until finally the organisms capable of surviving will be those which we call the simplest and least admirable. Of course "admirable" slipped in. If I were to take Dr. Waddington's definition seriously, I should not say that human beings were more 'admirable' or more 'valuable' organisms than the minute organisms that survive in the snow. I comfort myself for the step by remembering that after all Dr. Waddington does make the same value judgements as I have made.

Again, HItler, Goebbels, Himmler, Alfred Rosenberg, and Heydrich have also been produced by the evolutionary process. Suppose the Nazis won the war, and establish a new order in Europe, does that show the direction of the evolutionary process? If so, would that be good? This may seem to be an unfair criticism. But I do not think it is. Of course Dr. Waddington would say that it would not be good, because their triumph would not endure, and we must take long views, in the evolutionary process a million years is a short time—so that the 2000 years to which Hitler looks forward is a mere moment. But that answer will not do. For, if it be indeed true that *good means* 'in the direction of evolution', then, ex hypothesis, if Hitler & Co were the type evolved, then that type would be good. And the whole tenor of Dr. Waddingont's judgements show that he could not accept this. That he does himself evaluate the evolutionary process in non-biological terms is shown by his argument against Huxley's rejection of the cosmic process as constituting the standard ethical progress. He maintains (1) that the results of evolution cannot be adequately summarised as an increase in bloodiness etc.; and (2) that the course of evolution does not seem to us now to be 'so morally offensive that we cannot accept it'. This seems to me to contradict the assertion that good *is* the direction of the evolutionary process. If good be defined as that which is effective, i.e. that which is in the direction of evolution, what is the point of denying that the results of evolution are bloodthirsty? And what, on this view, can possibly be meant by saying that us *now*, as contrasted with Huxley, the course of evolution does not seem 'so morally offensive that we cannot accept it'? These two assertions suggest that Dr. Waddington does think it necessary to show that Huxley was mistaken in his estimate of the bloodthirst character of the evolutionary process. But this question of fact is irrelevant—i.e. the mistake Huxley made (if it was a mistake) about the nature of the evolutionary process can have nothing to do with Waddington's evolution. We must put this question to him: Suppose Huxley's estimate had been correct: Would it make sense to say that the evolution process was morally offensive? If Waddington replies that it would *not* make sense, then I wonder why he said it. If he replies that it would make sense, then it is clear that the evaluation *is* not derived from a contemplation of the evolutionary process.

I have dwelt on Dr. Waddington's account of good, because he argues from it that, in order to make a better world, it is necessary and sufficient that *science should rule*. Science—you will remember—is the only one of the five competitors that is capable of filling the gap after the Nazis have been defeated. [Turn back to page 1.]

The whole world, he agrees, 'is going totalitarian and nothing can stop it' (110) so let the dictators be scientists.

I am reminded of Plato, who, being a philosopher, thought that the ideal city could be attained only when philosophers were the rulers; also Leonard Woolf,[8] who, being an historian, thinks that a sane world would be one ruled by historians; and now Dr. Waddington, a geneticist, believes that a better world will be achieved only when science is *king*!

Apart from the comparisons between scientific statements and ethical statements based upon scientific statements—which I have already dealt with—there is the more facile comparison between *science* and scientists. It really is important to point out that *science can not* rule, not because science is too weak, or too stupid, or too neglected, but because it does not make sense to talk of science doing this, that,

[8] [[Leonard Woolf (1880–1969) was a British publisher, author, civil servant and political theorist who was married to the writer Virginia Woolf. Woolf, who also worked in the administration for the British government in Ceylon, is regarded as a representative of liberal internationalism. During his studies at Cambridge, he belonged to a group around G.E. Moore, J.M. Keynes and E.M. Forster (Cambridge Apostles) and was later a member of the Labor Party and the socialist Fabian Society.]]

or the other. It is *scientists* who act, *scientists* who make discoveries, *scientists* who teach us how much vitamin C we ought to consume. If Dr. Waddington had time to forsake his abstractions and to attempt to work out his views in detail, I hope we would take some note of the behaviour of scientists in their non-scientific activities. Do you seriously believe that scientists differ from other groups of professional men in always taking the *same* views concerning politics, art, religion? You cannot if you are acquainted with many scientists. Some would agree with Dr. Waddington in working for a totalitarian society; others will devote their efforts to stemming the movements that make for totalitarianism. Some scientists are socialists, others are communists, others are die hard tories of the bluest colour. It has not been my experience that the scientists on the staffs of universities always agree with one another in their educational policies, nor that they always bring to problems of management the dispassionate, literant, reasoned, temper that they, no doubt, display in their scientific work. Quite the contrary. If you protest, but, in that case the scientists are failing to be truly scientific, I will not deny that you are right. This indeed is precisely why I insist that we must remember that it is scientists who will have to rule, not *science*. In short, 'unscientific science' is a contradiction in terms; 'unscientific scientist carries out a scientific experiment' is a contradiction in terms; but 'unscientific Mr. X, who is a scientist, deciding questions of policy in non-scientific matters' is so far from being a contradiction in terms that, I am afraid, he is to be met on the staff of every university.

I agree entirely with Dr. Waddington that in the political social affairs of this world the scientist has an immensely important part to play in supplying information upon which sound policies must be based; but I entirely disagree that science covers the whole field, and emphatically deny that a scientist, in virtue of his prophecies in a special science is thereby fitted to rule the world.

Chapter 13
Men and Moral Principles

L. Susan Stebbing

The lecture that I have the honour of giving to-day is dedicated to the memory of Leonard Hobhouse.[1] I am glad to have known him, for Hobhouse was a philosopher whose philosophy informed his life. I count myself fortunate in having been associated with him for several years on University and other committees, during the work of which my respect for him was in no wise diminished but in fact increased. My academic colleagues, knowing how frequently human beings fall below themselves in committee-work, will realize that I desire to pay no small tribute to his personality and powers.

The topic I have chosen for discussion falls within the scope of moral philosophy. Hobhouse devoted much thought to this most difficult of philosophical studies. I realize very well the difficulties and my own inability to say anything clear about them. Questions properly belonging to moral philosophy are of fundamentally different types. (The word 'fundamentally' is here used in its strict sense.) The methods appropriate to the discussion of these different questions must accordingly be also of fundamentally different types. One set of questions are those arising from a consideration of the use of ethical language, i.e. of words which, when used in a sentence, enable us to recognize that an ethical statement is being made. Examples of such statements are 'Cruelty is evil'; 'Men ought not to be persecuted on account of their religion'; 'We ought to obey the laws of our country': 'It is good that a man should do what he thinks right'. I hope it will be agreed that these four statements

L. Susan Stebbing died before publication of this work was completed.

[1] [Originally published by Oxford University Press, in 1944 as the 13th L.T. Hobhouse Memorial Trust Lecture. Reproduced here by the permission of the Press, and by the kind permission of the copyright holders of Stebbing's materials, Alison Willott and Celia Pyke-Lees.]

L. S. Stebbing
1885–1943
London, UK

G. Schiemer (ed.), *The Legacy of the Vienna Circle*, Vienna Circle Institute Yearbook 30, https://doi.org/10.1007/978-3-031-80568-4_13

are all of them ethical statements, and that the question whether they are true or are false is a question falling within the scope of moral philosophy; and, further, that this question of truth and falsity cannot be answered unless we first know, in the case of each of these statements, what exactly it is that is stated. But with such questions as these I am not to-day concerned. I shall simply take it for granted that statements such as my four examples are frequently asserted, are sufficiently well understood as used in ordinary conversation, and can be, moreover, significantly used in ethical discussions. Further, I shall take it for granted that all of you could indefinitely extend the list of examples by adding others of the appropriate types. But although I here take all this for granted—since not everything can be discussed at once—I do not in the least deny that statements such as my four examples stand in need of analytical scrutiny. Analysis of this kind is, in my opinion, a proper part of moral philosophy. Analysis is, however, merely a propaedeutic to using what has been analysed. Certain contemporary philosophers seem contented to regard moral action only as providing occasions for ethical remarks and are interested in these remarks only in so far as they provide exercises in linguistic analysis; these exercises, treated as games requiring no small degree of specialized skill, are, like chess-problems to an ardent chess-player, inexhaustibly amusing but of no practical concern in the business of living. Moral philosophers, I contend, must be concerned with the ways in which men live—their ways of life which embody their ideals. I conceive that it falls within the proper province of moral philosophers to formulate ideals worth living for and to attempt to make clear principles which may afford guides for action. This is a task singularly difficult to fulfil.

It is very easy to become muddled at this point and even easier to say something that may be misunderstood just because one has failed to notice one's own muddles. The task I have assigned to moral philosophers requires us to be clear with regard to the relation between moral theory and moral practice. This relation is by no means easy to determine. Those who write and talk about the problems of moral philosophy—especially those who talk—may, in the main, be roughly divided into two classes: those who seek to develop a comprehensive system of morality and those who are concerned to preach men into living well. Fortunately, we need take no note of the latter class. Unfortunately, works on moral philosophy that do not fall into the class of disguised sermons tend to suffer from one or other of two grave defects: either they are trivial or they are abstract. I wish to consider briefly the second defect. In this attempt, in order to obviate certain common misunderstandings, I must discuss certain distinctions the elaboration of which may well seem tedious and also irrelevant. Whether tedious or not, these distinctions are certainly not irrelevant.

There is first the opposition of *theory* to *application* and of *theoretical* to *experimental*. This opposition can be stated by reference to the distinction between pure and applied mathematics and to the distinction between theoretical physics[2] and experimental physics. Pure mathematics is theoretical; it may also be said to be abstract. By 'abstract' in this context I understand abstraction from the physical world, from Nature, in the most comprehensive sense of the noun 'Nature'. Nothing

[2] [[Neurath's handwritten note: "a part of mathematics."]]

that happens in the world is relevant to the truth or falsity of any mathematical theorem. For this reason pure mathematics is sometimes said to be a conceptual scheme, or set of conceptual schemes; sometimes it is said to be sets of symbolic systems. The nature of the distinction between pure mathematics and applied mathematics can be most clearly seen in the case of geometry. Pure geometry (which, of course, includes what mathematicians call 'analytical geometry') contains many 'geometries', for example, projective geometry, Euclidean geometry, non-Euclidean geometry, *analysis situs*. Each of these geometries is a symbolic system, or, if you prefer a fashionable word, a *pattern* of mathematical concepts. In working out these systems, elaborating these patterns, nothing whatever is said about what happens in the world. Suppose, for instance, that I was teaching a class of schoolchildren some theorem in Euclidean geometry, for example, that the three interior angles of a triangle are together equal to two right angles. I should certainly draw a figure on the blackboard in order to help the children to follow the demonstration. The figure can be very roughly drawn (and, in my opinion, it is always desirable to draw figures, used for such a purpose, roughly—with lines that are perceptibly not straight, and so on). The rough drawing is permissible and desirable just because the drawn figure plays no part at all in the demonstration. Now suppose that a massive gravitating body or a violent dynamo is introduced into the room in which the class is being held. A physicist would say that 'the geometry of the room' is changed; there is a slight but definite distortion. This distortion, however, has nothing to do with the correctness of the demonstration of the theorem, any more than has the noise that may be created by the dynamo. Certainly the noise might prevent the children from hearing what was being said, but it would not affect their ability to understand the demonstration if they could hear it. I do not say that this analogy is exact, namely, that the use of the drawn figure as an aid to the children's understanding is exactly analogous to their need to hear what is being said, but I do assert that both are equally irrelevant to the validity of the demonstration. It is in this sense that geometry is abstract: no notice is taken out of the presence of the blackboard, the shape of the figure, or the chalk lines: physical and geographical facts, auditory and visual experiences, are all entirely irrelevant to mathematical demonstration.

In this sense of 'abstract', theoretical physics is less abstract, and it may correctly be said that in physics geometry is applied; this means that, in the domain of physics, geometrical statements are interpreted in terms of happenings occurring in the world.[3] Within limits (the precise determination of which can be neglected here) a physical theory is not independent of what happens in the world; whether a physical theory has application or not is for the most part settled by experiment. The final Test of a physical theory is to be found in the observation of sensible phenomena. Although sense-perceptions themselves are deliberately excluded as far as is possible from physical investigations, nevertheless physical science is concerned with the world that contains stars, planets, desks, and elephants falling down slippery

[3] A geometry can be interpreted without reference to the word, e.g. a geometry in which every point has co-ordinates which are real numbers can be said to *apply* to a system of sets of real numbers. This is an interpretation that remains within the domain of *pure* mathematics.

slopes. It is for this reason that some experimental physicists were profoundly shocked when Bertrand Russell told them that physics is abstract. In the precise sense in which pure mathematics is abstract, physics is not abstract at all. The only science which is abstract in this sense is pure logic, i.e. the theory of types of order, often called mathematical logic.

We do, however, sometimes so use the word 'abstract' that we can speak of arranging the sciences in order of decreasing abstractness. As is well known, the sciences have been thus arranged with mathematics at the highest level of abstractness, then physics, and so on, until we reach sociology at the lowest level of abstractness. In my opinion this is not an illuminating procedure; it obscures the fundamental difference between mathematics and logic, on the one hand, and the natural and social sciences, on the other hand. Using the terminology of Professor Whitehead, we can say that this order of decreasing abstractness begins with maximum abstraction from possibility; in other words, it relates to elaborateness of logical construction.

In this sense of 'abstract' moral philosophy is the least abstract of systematic studies; it is less abstract than economics, which is a science concerned with some only of the activities of men; it is less abstract than sociology, which is concerned (I believe) with social groupings and the various forms of social institutions. But to speak in these terms is likely to be more misleading than enlightening. Looked at from the point of view of the methods employed and the knowledge yielded by the different studies, the sciences cannot profitably be arranged in a hierarchical order in accordance with any quantitative principle of division. The distinction between the sciences that is relevant to the purpose of this discussion is to be found by considering the different sorts of situations with which they respectively deal. The geologist deals with geological situations, the economist deals with economic situations, the moral philosopher with ethical situations. I suggest that the various sciences and special studies should be defined in terms of the situations with which the various specialists respectively deal, and not—as is usually the case—conversely.

A situation is a state of affairs in which various circumstances are conjoined; hence, a situation has necessarily internal complexity. The situations which interest economists, for example, contain human beings engaged in business transactions. Such a situation may be called an 'economic situation'. Now an *instance* of an economic situation always contains much more than the circumstances taken note of by the economist. Take for instance the economic situation in which a man, say Mr. Churchill, is buying a box of cigars at a certain shop on a certain day. The situation contains Mr. Churchill, the shop and its furniture and stock, a shop-attendant handing over the box, perhaps other would-be buyers waiting and possibly looking on at this transaction, and not impossibly a vase of tulips on the counter. There is in this situation material for the biologist, the psychologist, the economist, and the moral philosopher—to mention these only. But the economist is not concerned with this definite and particular situation; it is of use to him only in so far as it enters as an item into, say, an abstract of the year's transactions in the tobacco trade. The economist, I understand, obtains his material from such sources of information as the *Statistical Abstract of the United Kingdom*. The information there given is arranged in such ways as to enable the economist to draw conclusions capable of being so formulated as to yield the sort of results which will enable him to answer the questions which interest him in his capacity of an economist. Certainly economists do not at all agree with regard to the nature of their science; there is, I understand, disagreement concerning the status of economic laws and the significance of these laws in relation to the effects to which the laws would commonly be said to refer. But, with some hesitation due to my ignorance, I assume that economists would agree that they are able to discover what may not incorrectly be described as 'economic laws', and that these laws can be applied to definite situations and can be used to predict economic phenomena.

If what I have been saying is correct, both the instance *of* an economic situation, which constitutes a single item in the original sources of information used by the economist to obtain his data, and the situation to which economic laws apply are extremely complex states of affairs. But *the instance* of an economic situation and the situation used by the economist as data for his laws and as test for their applicability are not of the same type. In other words, the phrase 'economic situation', used to refer to *This man is buying cigars*, is not used in the same sense exactly as 'economic situation' is used when we are speaking of the situations that constitute the data upon which the economist builds his science. The second is abstract relatively to the first. Certain features only in the complex situation existing at a particular time and place are selected by the economist, and isolated by him under the name 'economic phenomena', in order that he may study these isolated phenomena in a disconnexion from any elements irrelevant to his general purpose of establishing laws and making predictions. The economist thus deals with artificially constructed isolates, and it is in terms of *isolates* in this sense that economic laws are stated. It was an implicit recognition of the fact that laws logically require isolates that led to the concept of the *economic man*. But, perhaps unfortunately for economists, *Homo economicus* is not a term in an economic statement, still less a term in a mathematical theorem or a Latin exercise. 'The economic man' is a *man*; accordingly, the economist, so long as he is concentrating on matters belonging to his chosen science, must select for study certain activities of men which can only with difficulty be isolated from other activities. It is, however, unlikely (I think) that the economist, who is also a man, and not merely a scientist, will believe that he can wholly neglect the connexion between these economic activities and the other activities proper to men. If he does take note of these, he is widening the scope of economics, regarded as a science, and is in fair way to becoming a social philosopher. There are sound reasons (to which I shall return later) why men studying economics, i.e. economists, should also study something else, namely, moral philosophy and social philosophy; there are sound reasons none the less for insisting that, in the latter case, he *is* doing something else. Economic laws are logically independent of moral values.

I am aware that some economists would dispute my contention that the economist is doing something other than economics when he takes note of the worth of different ends and of possible choices between ends. It is certainly not my desire to presume to lay down limits within which the economist must keep on pain of not being an economist. Since these specialists disagree amongst themselves with regard to the scope and significance of their science, it would be foolish for an outsider to rush in where economists tread on one another's toes. Nor do I wish to suggest that any economists to-day would conceive economic man after the egoistic, hedonist fashion of, say, Jevons. The interests of economic man will be admitted to include not only the desire for profit but also the desire for leisure, for prestige, for independence, and so on. That is to say, characteristics such as these may properly be attributed to the economic man. Even if he has no heart, he is not a skeleton constructed out of one homogeneous material. The sole point I am concerned to make is to insist that in the formulation of economic laws moral values have no place.

In so far as economics is a science it is so because it is selective in the special way which permits of the formation of isolates. I maintain that a subject of study is a science only if isolates can be constructed and fruitfully used. Sciences differ in their degree of freedom in forming isolates, and I believe that this provides the sole justification for the hierarchical ordering of sciences in the order of decreasing abstractness. This does not in any way constitute a denial of my contention that no hierarchical order based upon a quantitative principle of division can be properly made, for it is not only the degree of freedom in forming isolates that is relevant. Sciences differ importantly in the nature of the isolates formed, in the extent to which their isolates can be rigidly maintained, and in the fruitfulness of the isolates for the purpose of establishing laws and enabling predictions to be made. The more rigidly the isolates can be maintained, the more exactly can laws be formulated. I suspect that economics has greater freedom in this respect than has sociology, but as I am not able to discover exactly what sociology includes, I hardly presume to offer an opinion on this point, although, unfortunately, that does not mean that I have no opinion. It is difficult to refrain at times from having opinions about matters in which one's ignorance is a disqualification for forming *reasonable* opinions. But one can at least keep them to oneself.

It follows from what I have been saying with regard to the part played by isolates in scientific laws that Politics (in the Aristotelian, not the parliamentary sense) can hardly give scope for the formulation of laws. I am aware that the late professor Collingwood claimed to have discovered 'three laws of politics',[4] but I do not think his claim can be admitted. His 'three laws' are not, I suggest, laws at all, precisely because the fundamental term *class* involved in the formulation of the laws has not been properly isolated and, indeed, the word 'class' is in fact used in two different senses in the verbal enunciation of the 'laws'. This failure to form an isolate adequate to the formulation of the laws makes havoc of the claim that *laws of politics* have been stated. This outcome I do not find surprising. Politics, I maintain, cannot profitably be regarded as a science in the sense in which economics is, or aspires to be, a science; on the contrary, politics is a branch of philosophy. The name 'moral philosophy' is indeed in no sense a misnomer except for the unfortunate ambiguity that attaches to the word 'moral'. No one would now, I suppose, wish to restrict the

---

[4] In his Hobhouse Memorial Trust Lecture, entitled 'The Three Laws of Politics', and in his book *New Leviathan*.

application of the word to virtuous sexual conduct; nevertheless, it is very commonly used less widely than the word 'ethical'; it tends to be limited in application to judgements on persons and judgements about conduct. I do not myself find it desirable so to restrict the usage of the word, but, to avoid misunderstanding, I shall use the word 'ethical' in the wider sense wherever it seems probable that 'moral' would be too narrowly interpreted. Moral philosophy is not confined to the evaluation and criticism of conduct; it is concerned with good and evil wherever they may be found. In short, moral philosophers are concerned with ethical situations; an ethical situation is a complex state of affairs in which good or evil is present.[5] It follows that every ethical situation contains at least one sentient being.

It is important to insist that ethical situations are for the most part situations of extreme internal complexity containing elements of ultimately diverse types; consequently, judgements of evaluation with regard to them are also of diverse types. Lack of time prevents me from discussing properly the difficult and important questions that arise with regard to this diversity of types—questions that are in my opinion far too little noticed by moral philosophers. In spite of the need for brevity I must call attention to the fact that ethical judgements differ in logical type, and it is this that constitutes the main reason why the application of moral theory to practice is beset with difficulties.

Consider, for example, the pair of ethical judgements: *Pain is evil; Cruelty is evil.* These are diverse types. *Pain is evil* I shall call a simple, or basic, ethical judgement in the quite precise sense that the truth of the judgement is not based upon any other ethical judgement that is more logically simple, but it is itself a basis for other ethical judgements. *Cruelty is evil* is not thus simple or basic; its truth is based upon the judgement *Pain is evil* and follows from this judgement together with the intrinsic nature of *cruelty* as being enjoyment in witnessing or inflicting pain upon other sentient beings. I should myself say that it follows from the judgement that *Cruelty is evil* that under all circumstances and in all conditions to act cruelly is evil. This follows, however, from the total judgement, not merely from the fact that what is predicted of something is that it is *evil.* It is the intrinsic nature of cruelty that logically necessitates both that *Cruelty is evil* and also that to act cruelly is evil. It does not at all follow from the judgement *Pain is evil* that deliberately to inflict pain is to act evilly, nor that a situation in which pain is present is necessarily evil on the whole. The pair of judgements I have cited as examples are clearly diverse in logical type.

A judgement of a third type has been mentioned in what I have just said: *To act cruelly is evil.* This judgement could not be true unless the judgement *Cruelty is evil* were true; and cruelty could not be evil unless it were true that *Pain is evil.* I insist upon this point, which had, indeed, seemed to me the merest platitude until I discovered that it has been implicitly and explicitly denied. It is denied by anyone who maintains that there is nothing that is intrinsically evil except the evil will, and by Kant who maintained (although not consistently) that there is nothing good except

[5] [[Neurath's handwritten note: "the words 'good' and 'evil' used".]]

the good will. Those who maintain that only sin is evil and that only doing one's duty—or acting in obedience to God's will—is good are logically bound to deny that, for instance, bodily pain is evil, that enjoying music is good, that the admiring contemplation of the beauty of Einstein's relativity theory is good. Unfortunately, this is just what is denied. Hence, it is important for me to assert the contrary view with all the emphasis I can command. Using the term 'moral values' to describe excellences of character and conduct, I assert that moral values depend upon non-moral values; if nothing except these excellences were good, then it would not make sense to say that *they* were good.

Certainly it does not follow from the fact that enjoying music is good that it is always good that I should be enjoying music; whether this is good or not depends upon the whole situation in which the enjoyment is an element. Nor does it follow that it is good that everyone should enjoy music. My point is that a judgement of the form *It is good that so-and-so* relates necessarily to a particular ethical situation. No inspection of the elements of good and of evil in that situation, separately considered as respectively good or evil, view by itself suffice to enable us to determine whether the situation is itself good, or evil, on the whole. We are, at least in the case of ordinarily complex ethical situations, confronted with a variety of judgements of utterly diverse logical types. Owing to the diversity it is extremely difficult to select even relatively stable isolates in terms of which judgements of good and evil can significantly be made with regard to sets of ethical situations as wholes. Consequently, it is rarely possible to formulate any theoretical principles from which moral rules can be derived such that these rules ought to be observed by everyone, or even by most people.

It is not to be disputed that we make generalizations in ethical terms. I have already cited one, namely, *Enjoying music is good.* Other examples are, *Promises ought to be kept, We ought to speak the truth.* These are commonly designated 'moral rules'. Consider the rule, *Promises ought to be kept.* In any situation in which a promise has been made someone has accepted an obligation to act in a certain way on some expected future occasion. Now, a generalization to the effect that such an obligation must be met—a generalization with regard to conduct—may seem a little comparable to the statement of an economic law, so far as the form of generalization is concerned and neglecting the very considerable difference in subject-matter which involves an *ought* instead of an *is*; for, in the interpretation of an economic law it is presupposed that certain conjunctions of circumstances may render the law inapplicable and the prediction it contains falsified. But there is in fact *nothing* comparable in the two cases. An economic law simply states that under such and such combinations of circumstances such and such will happen. Relatively to *these* combinations of circumstances the law can be completely stated. The case is quite otherwise with the moral rule. The combination of circumstances which determine, when the time for keeping the promise comes, whether there is anything in this situation that is ethically relevant necessitating that the promise ought to be broken cannot be generalized in any useful manner. The *whole* force of the rule is to impose upon us the burden of justifying the breaking of the promise provided that elements in the situation necessitate the judgement, *In this case the promise ought to be broken*. The rule does no more than tell us that to keep a promise is good. The same is true of the rule *We ought to speak the truth*, and any other moral rule.

The difference between an economic law and a moral rule that is important for this discussion depends upon the way in which each is related to its respective situation, and this is in turn dependent upon the different nature of the abstraction involved in speaking of these two kinds of situations. It is because an ethical situation is relatively non-abstract that a moral rule cannot be applied after the fashion of an economic law; i.e. cannot be applied *in the same sense of* 'apply'. This point may perhaps be made clearer by means of a definite example. Suppose that *I* (i.e. any *one* person) am in a situation in which I must reply to a question asking for information in my possession. I acknowledge—it is assumed—the abstract moral rule *Speak the truth*. But the description 'I am now speaking the truth' is never a complete description of any situation at all; nor does the description ever include all that is ethically relevant for justifying obedience to the rule. It may be that in *now* speaking the truth I am also wounding a friend to his grievous hurt, or betraying my comrades, or breaking a promise, or destroying someone's illusion. Indeed it is by no means impossible that it should be two or more of these together. Now the rule is an isolate: *Speak the truth*—just that and nothing more. But, as we have just seen, there must necessarily be something more. By making an isolate we obtain the rule; in thus isolating we abstract from circumstances in the situation that alone make it a duty to speak the truth (if so be that *here* the rule ought to be followed) no less than we abstract from circumstances in the situation which alone make it a duty to lie (if so be that *there* the rule ought not to be followed). The certainty of the rule comes from our forgetting that it is impossible for an action to be merely a-speaking-of-the-truth.

We cannot get rid of this difficulty by trying to form wider isolates in terms of which to state our moral rules. *Speak the truth provided that you do not thereby break a promise*; clearly this has no universal validity. *Speak the truth provided that you do not thereby wound your friend*; this will serve us no better. We form isolates, we combine and recombine them; the result reminds us of copybook maxims and of certain 'talks' heard on the wireless. 'So snappy!' we may even be tempted to say, 'so true!' But when we are called upon to act, how feeble seem these 'talks', how useless these maxims!

The point which I have tried to make clear by means of the example of truth-speaking is that what may seem to be extraneous circumstances are really factors integral to the situation as an ethical situation, and are consequently relevant to the application of a principle or the following of a moral rule of action. I think this contention may be more readily admitted if we look at this problem from another point of view, namely, that of evaluating someone's conduct instead of asking what rule we should follow.

Consider these two examples: (1) I give a thirsty child a drink of cold water; it costs me no effort; there is plenty to spare, and the tap is at hand. (2) Two torpedoed sailors on a raft at sea are both in great pain from thirst; one, at the point of dying, calls for water; the other, whose need is no less great, gives his companion the last drop to drink. We judge these actions very differently. The first has at most but negligible moral worth; the second has indubitable moral worth. Now it is a mistake to say that we are judging the same action in each case but the circumstances are so different that the actions differ in moreover worth. On the contrary, they are entirely different actions. In (1) the action is *giving water to someone when I have no need of it myself and without my suffering effort or loss*. In (2) the action is *giving water to someone who needs it when I need it no less myself and my giving the water is not without suffering and loss to myself*. Nothing less than this suffices to describe the action that is judged. All that the actions have in common is *giving water*, a characteristic which does not itself entail any ethical property.

It seems then that we must be on guard against isolating situations no less than isolating characteristics of actions. What we have to take account of is a whole mode of acting; ethical principles have significance only in so far as they are closely related to a certain set of actions which proceed from a spirit that inspires a particular way of life.

If what I have been saying is in the main correct, it follows that the methods used by a moral philosopher must be entirely different from the methods used by the natural and social scientists for the sake of introducing order into the complex phenomena with which they respectively deal. In no illuminating sense can it be maintained that there is in principle, even if we have not been able to discover it, *a* moral theory that can be directly applied to practice. The ordinary methods of deduction are inapplicable. I here take for granted, because I have not time to discuss their assumptions, that scientists who think they have found the all-sufficient principles of ethics in the laws of biological evolution are sheerly mistaken. Moral philosophy, I repeat, is not a science.

What method, then, can we use? None that has been labelled or described in detail; no method that a self-respecting pure logician can find joy in pursuing. When I think of the neat, trim, tidy problems of logic and philosophy of science and then turn to these sprawling, tangled problems of moral philosophy I shudder with distaste and long for a logically elegant procedure. Most philosophers, I think, must have felt as I do. How else can we explain the fact that so many of them have been more abstract than was necessary? Some, for example Kant, have tried to find a single ethical principle, a single moral motive; others have sought a single good, for example, the Hedonists; others again have taken refuge in a huge abstraction, for example, self-realization. All alike have failed. Often they have illuminated some problems, partly solved some difficulty, or have laid bare some of the essentials, but always, *in spite of*, not *because of*, the method they have used. It is a mistake to suppose that pure hearts or clear intellects suffice for the task. Out of the mouths of babes and sucklings we cannot hear any moral truths relevant to our lives, for these innocents do not know what it is like to live. We cannot by merely thinking, however brilliantly, solve any problems at all, for thought alone deals only with abstracta. It is as useless as it is profoundly unphilosophical to formulate moral ideals without taking note of the manner of men we are. Whatever may be the case with politicians making week-end speeches in time of war, philosophers cannot afford to ignore the conditions of the problems set by the situations in which we live.

At the outset it is well to remind ourselves that we cannot, by taking thought, make everything clear; at most we can but make some things clearer. There will inevitably be dark patches and ragged edges despite our utmost efforts. Lack of time prevents me from doing more than hint at the procedure that seems to me to be appropriate. Here I have found most help from Aristotle, apart from one big blunder, and from G. E. Moore.

I will use in illustration the moral rule I have already taken as an example, *Speak the truth*. We saw that it was idle to seek a universally applicable guide in this rule; there is no universal justification for telling a lie. Only by looking at the given situation is it possible to know whether a lie ought to be told. Suppose such a situation arises and the lie is told. In telling the lie, then, I acted rightly. This does not, however, in the least diminish the evilness of lying; *that the lie had to be told* is evil. Can we formulate an ideal way of life in which no such evil situation is found? This question is unclear and admits of two contrary answers. Certainly if our imagination is lively we can picture to ourselves a state of affairs in which no one is so weak in body or so broken in spirit that he cannot bear the truth, in which no promise must be broken unless a lie be told, in which no one has to choose between betraying his comrades and telling a lie. Such a picture is a fairy tale. Our concern, however, is with actual men, living here and now, in this world. Accordingly, if the ideal way of life we said before as an aim is to be an ideal that we can now begin to put in practice, we must not postulate conditions that are unrealizable and thus solve our problems by denying that they exist. People often seem to think that it would be possible, *provided only* that we were set on doing our duty, to leave out all the evil elements in the existing situation and retain all the good ones. This is a mistake, of which you might be able to convince yourselves if you would seriously try the experiment of continuing to live in the community in which you do now actually live and of going about your ordinary affairs and then, for one month, practice never to make a petty evasion or to express more liking for something or someone than you really feel, or to deal tactfully with a situation in which tact is incompatible with truthfulness. To take a tiny illustration—for sometimes we can more easily discern small letters than large. It is customary in the community in which we live after a public lecture for the friends of the lecturer or for the chairman to say something pleasant and approving to the lecturer so that he may go away at peace with himself. Suppose the truth is rigidly adhered to so that, after a feeble lecture, the lecturer is left without this needed comfort; something is lost. I do not, of course, want to assert that in *this* example the lie, or the petty evasion, would be a justified lie. What I am asserting is that the cutting out of the evil of lying is not unaccompanied with the loss of something it might, under easily imaginable circumstances, be good that a man should have, though unearned.

Pacifism provides a more serious and larger example. It is larger because any pacifist view that is worth taking note of is itself an ideal; i.e. a conception of a way of living that is judged to be good; it is a positive conception, not a mere negation. Pacifists, I believe, usually assume that it is possible here and now, in this England, to achieve the ideal of pacifism without surrendering anything that is supremely worth while. They cannot, however, avoid the choice between evils. To take the contemporary situation as an example. I should contend that we had the alternatives of allowing Hitler's way of life to prevail at least in many countries on the Continent and perhaps also in this country or going to war with Germany. Pacifists, or at least some of them, maintain that neither of these evils had to be accepted provided that *we* (say, the British) in 1938, or 1939, had definitely and openly adopted the Christian way of life. Even if this contention were true (and I do not myself believe it is) it seems to me an evasion of the problem, for the solution is stated in terms of an hypothesis which we know to be unrealizable, and hence inapplicable to the situation in which we are called upon to act. In our community the vast majority of people simply do not believe in the practicability, and many do not believe in the desirability, of the Christian way of life as understood by a Christian Pacifist. To argue from an hypothesis that if people were to live in a way in which we know they neither do nor can immediately live, then certain evils could be wholly eliminated without accompanying loss of what is good, is assuredly not to show *how* we can realize those goods and cut out those evils; it is to live in a fool's paradise.[6]

This line of argument is not, however, the most pertinent to my purpose. Let it be granted, for a moment, that a sufficient minority in this country had been practising Christian Pacifists during the last ten years to prevent our declaring war. Their realizing of their ideal would in no way have hindered the effective domination of European peoples by Hitler's ideal. I need not elaborate this contention. My point is that in that case the pacifist country would have chosen between evils and would have decided that the Nazification of German children is a lesser evil than attempting to free both Germans and the rest of Europe from being indoctrinated with the Nazi ideal. The evil would not have been cut out from the whole situation leaving only what is good; at best, one country would have avoided the evils of war and have borne witness to a way of life that could not be followed universally.

It seems, then, that it is not only philosophers, loving 'unity' and 'system' so greatly that they will seek them at all costs, even at the cost of truth, who abstract too much and make for themselves unfruitful isolates; men in love with a noble ideal are tempted to shut their eyes to the evils others promote and suffer whilst they devote themselves to attempting to attain an ideal the achieving of which is incompatible with realizing certain goods which nevertheless they also accept and then ignore. Their error is in part a refusal to face the fact that there are incompatible goods between which a choice must be made. To make this refusal is to be guilty of intellectual dishonesty.

It may be objected that what I have been saying logically implies that we should compromise with our ideals. Such an objection would be mistaken. Ideals fit for men are not Utopian schemes; they are for putting into practice. What I urge is that

[6]Cf. 'No, friend, you do not beat me: hearken why! / The common problem, yours, mine, everyone's, / Is – not to fancy what were fair in life / Provided it could be, – but, finding first / What may be, then find how to make it fair / Up to our means: a very different thing! / No abstract intellectual plan of life / Quite irrespective of life's plainest laws, / But one, a man, who is man and nothing more, / May lead within a world which (by your leave) / Is Rome, or London, not Fool's-paradise.' (R. Browning, *Bishop Blougram's Apology*)

ideals are to be taken seriously as aims to be striven for. This means that we must take account of the situation within which these ideals are to be achieved, if at all. It is, therefore, relevant to inquire what the given situation permits to be done. An ideal involves a set of interconnected principles regarded as providing guides for action. A principle that is so much divorced from what the situation permits that we cannot see *at all* how the ideal can be exemplified in practice is at best an empty abstraction, at the worst a cover for hypocrisy or daydreaming.

I have not left myself time to say in sufficient detail how ideals can be used to provide guiding principles. I can only indicate in brief a possible procedure.

A difficulty, which my reference to Christian Pacifism was intended to bring out, arises from the fact that our ideals are apt to be partial—to concentrate on some great goods and to abstract, in consequence, from certain essential factors in the total situation and thus to ignore the circumstances in which men actually carry on their daily activities. In thinking about this problem I have found some help from a passage in Professor Tawney's *Religion and the Rise of Capitalism*. Speaking of the difficulty of applying the same moral standard to 'decisions which affect large masses of men as to those in which only individuals are involved', Professor Tawney puts the dilemma: 'To argue, in the manner of Machiavelli, that there is one rule for business and another for private life, is to open a door to an orgy of unscrupulousness before which the mind recoils. To argue that there is no difference at all, is to lay down a principle few men who have faced the difficulty in practice will be prepared to endorse as of invariable application, and incidentally to expose the idea of morality itself to discredit by subjecting it to an almost intolerable strain.'[7] Now it is sometimes the case that a set of men engaged in earning their livelihood can adopt principles of conduct binding upon them all and of such a kind as to conduce to the best interests of those whom they serve. I take as an example the medical profession. The main factor in medical practice is the nature of the relation between doctor and patient. If the doctor is to do his work he must be trusted. This does not mean merely that the doctor must be kindly and personally reliable and punctilious in performing his engagements; he must be reliable in his *medical knowledge and skill*; that is to say, he must have special knowledge and skill. Further, the patient on his side must be able to distinguish the skilful and well-informed medical man from a quack. The urgent needs of both doctor and patient have led, in the course of years, to the development of a professional code the acceptance of which is in the best interests of both the parties concerned. The Oath of Hippocrates gives a rough indication of what the code enjoins. It is, within its limited domain, a high ideal, but it is not an ideal that is essentially unpractical; on the contrary, it is in the main practiced.

I take this merely as an illustration of procedure. Had I time I might also have considered the teaching profession, or even the Law. The point I wish to illustrate is that in formulating an ideal we are necessarily thinking about a way of living, not about a set of principles which can be neatly stated but which do not in themselves

---

[7] Op. cit., p. 183.

provide us with any hint with regard to the way in which they can be exemplified in our lives. Our method, then, in my opinion, should not imitate that followed by most moral philosophers. For, unless I am greatly mistaken, these philosophers have been mainly concerned to provide a list of goods, or a set of principles—or, worse still, a single good or a single principle—and have held this up to our admiring contemplation without asking *what it would be like* if everyone pursued this good or these goods, and accepted these principles, or this principle. They have assumed that life would go on much as before, *only*—everyone would do his duty and be good, would have merit and would be happy, or at least worthy of happiness. They have not tried to see how in a *given* social situation these principles *could* be put into practice. The method I should like to see tried would involve much less abstraction. First, we should consider ideals, or ways of life, such as people profess to live by and not merely think about. Thus we might try to work out in considerable detail, and with a wealth of illustrations drawn from the examination of particular situations, various ideals which claim adherents in the contemporary world, e.g. the Nazi ideal, the Russian ideal, the Christian ideal, the democratic ideal, the capitalist ideal. Secondly, we should look carefully at the illustrations we have used with a view to seeing what of good and what of evil each exemplifies. For certainly no one of them will exemplify only what is good or only what is evil. There are incompatible goods no less than there are incompatible evils. Each ideal, if it were put into practice, would promote, or at least tend to promote, the development of a certain type of man and would be inimical to the development of some other types. Thirdly, then, we must compare these types of men seeking to discover which holds the greatest promise of developing into this sort of person we believe a human being is capable of becoming. This means that we could not neglect any of the activities of men; we should have to take note of a man's business transactions, for instance, and to evaluate his behaviour in these transactions in the light of his other activities and of the principles which have been found applicable to these. To give a brief illustration: The medical code forbids a doctor to keep secret a cure for disease that he has discovered; *all* his special knowledge must be made available for the use of the whole profession. But—so I understand—an entirely different practice would be regarded as permissible in industrial affairs. Thus suppose, for instance, an invention enabling gramophone records to be put on the market much more cheaply were offered to a manufacturing firm. It would be in conformity with what is so only called 'the ethics of business' for the firm to buy the invention and then not to use it if, by doing so, greater profits might be made.[8]

[8] Prof. Hayek assures me that this hypothetical case involves a fallacy; it would not occur in practice, he urges, since it would not be to the manufacturer's interest to buy and then put into cold storage any invention which in the public interest should be used. I am not especially concerned to deny this statement, which is a question of fact, viz. whether the suppression of an industrial invention ever would, in the long run, be your advantage in increasing the manufacturer's profits. My point is that, if it *were* to endanger his profits, the manufacturer would desire to suppress the invention. More generally, the point I wish to illustrate is that the manufacturer's main concern is increased profits; his duty, in his official capacity, is commonly regarded as the duty of increasing profits and safeguarding the shareholders. In a competitive society this is a condition of success in running a business. The medical man, I assume, also want his practice to 'pay'; that is, he wants a

It must not be assumed that, in using these illustrations, I am hastily judging that whilst medical practice is good in this respect the manufacturer's practice is evil. Still less do I wish to suggest that a manufacturer could rightly act otherwise in the situation in which he is placed in our society. On the contrary. What I have been suggesting is that the difference in practice should be considered from the point of view of the way of life which consistent following of these principles would respectively entail. It would be necessary to take into account all the relevant conditions of the industrial system which necessitate that the manufacturer should act as, in my illustration, he would act in order to secure the ends for which the system is organized, namely, profits in a competitive society. This suggestion that he should act otherwise, regarding the ends the organized industry is designed to secure, would ignore the other evils that might thereby be introduced. My contention throughout is that it is absurd to take a single isolated feature and then pass on unconditional judgement of evil upon it and straightway demand that it be at once eliminated. For, in eliminating it, something good may also be destroyed. If we have in mind the contemporary world it is easy to see my reason for saying at the outset that men studying economics should widen their interests and also study moral and political philosophy. For it cannot be denied that some of the evils we all admit and some of the goods are intimately bound up with a given industrial system. In passing our judgements of evil we may find that what we have to condemn is whole system. Thus to condemn is futile unless we know what to put in its place.

It is not logically less absurd to confine our attention to an isolated feature that is unquestionably good and then to pass an unconditional judgement that this good must be secured at all costs. The cost has always to be taken into account. What moral philosophers and earnest reformers sometimes forget is that the cost of something good may be something that is also good; our choice is not invariably a choice between good and evil but between incompatible goods. No goods are incompatible in isolation; the incompatibility arises from the situations in which these goods have to be realized.

The procedure I have so briefly, and wholly inadequately, indicated is clearly untidy. To follow it would take a long time; it would require the co-operation of many experts, and, even at its best, it could not yield conclusions universally acceptable. Nevertheless, it may well be doubted whether any speedier method can fruitfully be employed, if we are serious in believing that moral principles are principles for guiding the lives of actual men. The determination and clarification of these principles is not a matter to be settled by definitions. Our problems can be solved, if at all, only after careful consideration of the complex situations in which men actually live. There is no good reason to suppose that *one* way of life, one clearly stated ideal, is appropriate to all stages of human development and to all sorts of conditions of societies.

---

wage for his work. But the nature of his work has necessitated the kind of professional code to which I have referred. In business affairs the interests of the consumer enter only insofar as they affect finding markets. In the one case *profits*, in the other *service*, is the main consideration. Hence, I do not think that Professor Hayek's objection invalidates my illustration.

Part III
Reviews

Chapter 14
Review: Alexander Linsbichler, *Viel mehr als nur Ökonomie: Köpfe und Ideen der Österreichischen Schule der Nationalökonomie*, Böhlau/Brill 2022

Reinhard Neck

Over the last few decades, interest in the Austrian School of Economics has increased in the United States and around the globe. This has both academic and political reasons. From the academic perspective, the sequence of economic crises (the financial and economic crisis or "Great Recession", the COVID-19 pandemic) has raised doubts about the ability of mainstream ("orthodox") economic theory to provide adequate explanations of such phenomena and valid recommendations to politicians to deal with them. Looking for alternative (heterodox) theories is a natural reaction in such a situation, and besides Marxist, Neo- and Post-Keynesian, Neo-Ricardian, Institutionalist and other approaches, Austrian Economics has also been proposed as an alternative, despite (or perhaps: because of) the divergent political conceptions behind these different "unorthodox" schools.

The book reviewed here gives a popular scientific overview of the history of the Austrian School of Economics, with a focus on the lives and works of ten of its members, all but one of them born in (what was at that time) Austria, although fewer than half of them flourished and died in that country. This points to the fate of Austrian Economics, to what Karen Vaughn called "the migration of a tradition".[1] This was due to the political history of Austria: its loss of a hegemonial position in an empire after the demise of the Habsburg Monarchy after World War I, the anti-liberal and antisemitic (both Christian and Nazi inspired) atmosphere in the inter-war period, and eventually—and most threateningly—the persecution of "non-Aryan" people (and especially intellectuals and scientists) after Austria's

[1] Karen I. Vaughn, *Austrian Economics in America. The Migration of a Tradition*. Cambridge, UK: Cambridge University Press 1994.

R. Neck (✉)
Department of Economics, University of Klagenfurt, Klagenfurt, Austria
e-mail: Reinhard.Neck@aau.at

G. Schiemer (ed.), *The Legacy of the Vienna Circle*, Vienna Circle Institute Yearbook 30, https://doi.org/10.1007/978-3-031-80568-4_14

occupation (the "Anschluss") by Nazi Germany. Members of the Austrian School, whether "non-Aryans" or not, were targeted by the Nazis and had to emigrate. Their discrimination started already in the First Republic of Austria, when, for instance, Fritz Machlup could not obtain a "habilitation" (a postdoctoral thesis in the German-speaking world) and Ludwig von Mises, although the most diligent and productive economist of his time, was denied a position as professor at the University of Vienna.

In spite of the adverse climate in those times in Austria, members of the Austrian Economics camp and other intellectuals gathered in cafés or non-university institutions, such as the Chamber of Commerce of Vienna, to discuss papers and ideas of the participants in a liberal and tolerant manner. The Mises Circle was one of them, which, after Mises's emigration to the US, continued to some extent in the Mises seminar in New York City, although during his lifetime it was not much more than a sect within the—then overwhelmingly Keynesian—community of US economists. Nevertheless, in view of the libertarian positions of this group, it became influential with the advent of "neoliberalism" during the Reagan and following US administrations, and nowadays these "Neo-Austrians" are a non-negligible group of academics and especially business people. Mises and Hayek institutes were and are continually being founded and funded by rich and devoted sponsors and are spreading all over the free world. Unfortunately, the ideologies of many of these groups do not have much in common with the open and tolerant positions of the Austrians in Austria before World War II, as they support right-wing and even authoritarian politicians, such as Donald Trump in the US and Javier Milei in Argentina, for instance.

The book by Linsbichler provides an overview of the thinking of "Austrians from Austria", from Carl Menger, who is rightly regarded as the founder of the School, to the last generation of those with an Austrian background in the years before World War II. Nowadays in Austria, there are only very few scholars who can be counted as Austrian School economists or at least with a genuine interest in that school. This is a pity as their knowledge and education within the Austrian historical tradition could serve to infuse some ideas into the academic discourse not present now in the US or elsewhere. For instance, it is true (but often denied by ideology-driven "Neo-Austrians") that Carl Menger was not a "paleo-liberal" (to use a term coined by the former long-term holder of Menger's chair in Vienna, Erich W. Streissler) but could be seen as being close to a "liberal" in the US sense of the term, not denying the need for social policy and heavily attacking the conservatives of his time (together with his most famous student, Crown Prince Rudolph, albeit anonymously and without success). The Austrian School was much broader then than what runs under this label today, being open to ideas from other disciplines, such as philosophy, logic, mathematics, jurisprudence, and the humanities.

It is a great merit of Linsbichler's book to emphasize the interdisciplinary nature of the Austrian School of Economics, which is already prominent in its title: "much more than economics". After a short general introduction, Linsbichler presents the biographies of Carl Menger, Friedrich von Wieser, Eugen von Böhm-Bawerk, Ludwig von Mises, Joseph Alois Schumpeter, Martha Steffy Browne, Friedrich August von Hayek, Gottfried Haberler, Fritz Machlup, and Oskar Morgenstern. Probably only Mises and Hayek would be recognized as Austrian School economists by some "neo-Austrians", which demonstrates the narrowing of their perspective. As Linsbichler shows, even Schumpeter, who admired both Walras and Marx, and Morgenstern, the co-founder of game theory and other mathematical theories, contributed to and were influenced by the Austrian School's ideas and theories. Another merit of the book is the chapter about Steffy Browne, a very underrated economist who wrote an excellent treatise on the theory of economic policy as her habilitation, which could not succeed in Austria as she was both "non-Aryan" and female. In general, the inclusive approach to the question of "who is/was an Austrian School economist" in Linsbichler's book is highly laudable.

The popular science character of the book, which is aimed at a general audience, has its advantages and disadvantages. The former comes from the colloquial language, which makes the book an easy read. It includes many stories about the "heroes" of the school, without falling into the trap of becoming hagiographic. For instance, neither Wieser's later swing towards authoritarianism nor Mises's stubbornness are concealed. The drawback of the light style of the book comes when the stories degenerate into gossip whose origin is unclear. This is the price to be paid by the absence of a scientific apparatus, which may be an asset for the general reader but a nuisance for the expert. Moreover, although the author works hard to explain the theories of the biographees and their colleagues in plain language, this endeavour sometimes leads to oversimplifications.

To summarize, Linsbichler's book makes a valuable contribution to the literature about the Austrian School of Economics, providing a different (Austrian or European) perspective on that school for the German-language reader. Its translation into English is highly desirable in order to acquaint readers from the Anglo-Saxon world with the origins and development of the school, which are often not present or biased in treatises on the other side of the Atlantic. For such a translation, and for a possible second edition, at least a minimal amount of information about the sources of some of the stories told would contribute to the academic reputation of the book and its author without diminishing the pleasure of prospective readers.

Full Title:

Alexander Linsbichler, *Viel mehr als nur Ökonomie: Köpfe und Ideen der Österreichischen Schule der Nationalökonomie*, Vienna: Böhlau/Brill 2022. 278 pp. ISBN: 978-3205215691.

Chapter 15
Review: Kurt Gödel, *Philosophische Notizbücher/Philosophical Notebooks*. Volumes III–V. Edited by Eva-Maria Engelen, De Gruyter 2021/2023

Karl Sigmund

It is a topic of romantic tales; a glacier melts, and human remains return to day-light, uncannily preserved by the slow-moving ice, bringing some of the remote past back with them. Much the same is happening with Kurt Gödel's thoughts. Many decades ago, he jostled them down in Gabelsberger shorthand, notebook after notebook, strictly for his own use. Preserved in the archives of Princeton, his pencil scribblings slowly faded away, year after year, seemingly lost for posterity, illegible and undecipherable. Today, however, several groups of dedicated scholars bring Gödel's writing back to life, transcribe it, comment it, translate it. Never has Gödel published more than now.

The volumes III to V of his philosophical notebooks were written between the late summer of 1940 and the late spring of 1942. After his dramatic escape from the Third Reich by means of Trans-Siberian railway and Trans-Pacific steamer, Gödel had managed to reach the safe haven of the Institute for Advanced Study in Princeton. He would never leave it again—but this, of course, he did not know. When he wrote notebooks III to V, he was a refugee from Europe, and eventually even an "enemy alien", with an untenured future, and a past crammed with intellectual triumphs and psychic torments.

The first two volumes of Gödel's "Max Phil" notebooks, written in Vienna, had been concerned to a large extent with maxims for organizing his life and optimizing his scientific work. Such maxims still show up in the following volumes, but less and less so. A large part is on philosophy, psychology and theology, with frequent excursions to the foundations of mathematics, the whole interspersed with a few personal remarks. Since the notes were never meant to be read by anyone else but Gödel himself, the overall effect is a bewildering mix, to say the least—a kind of intellectual bouillabaisse. The editor Eva-Maria Engelen has done a superb job in

K. Sigmund (✉)
University of Vienna, Vienna, Austria
e-mail: karl.sigmund@univie.ac.at

G. Schiemer (ed.), *The Legacy of the Vienna Circle*, Vienna Circle Institute Yearbook 30, https://doi.org/10.1007/978-3-031-80568-4_15

sorting out the various ingredients and threats, with explanatory footnotes on almost every page and a wealth of cross-references, but it is obvious that generations of Gödel scholars will continue mining and re-interpreting the text, and still remain baffled.

Young Gödel had been an official member of the Vienna Circle, but his mindset was utterly different from the "Scientific Worldview" of logical empiricism. It is much closer to Leibniz, Newton, or the ancient Greek philosophers. His "essential philosophers who need to be studied" are Plato, Aristotle, Thomas, Descartes, Leibniz, Locke, Kant, Hegel, Plotinus—and here, the list stops (volume IV, page 188).

In truth, many of Gödel's notes are darker than fragments of the Pre-Socratics. Enlightenment, Marxism, Darwinism, or language analysis are conspicuously absent, or show up, at best, as trace elements. The Bible, by contrast, is frequently referred to. "Perhaps one should learn the Bible by heart, even where one does not understand it, just as one first repeats sounds before learning a language (v. III, p. 268)." Some of Gödel's remarks have a distinctly occult ring. He asks himself: "Is there also a sin of inanimate matter? (IV, 202)" and he notes a few lines later: "Color of God = black (because he is nothing in this world)."

The devil occupies a considerable place in Gödel's notebooks, and corresponds to "an irrefutable wrong system (or the empty set) (III, 247)". God, accordingly, is the universal set. Gödel must have been aware of the underlying irony: Indeed, in present day mathematics, the empty set is the only set that is explicitly given (by axiom), whereas the universal set is not a consistent notion: it immediately leads to contradictions.

No one, however, who is cognizant of mathematics will underestimate the power of the empty set, and it is only to be expected, therefore, that the devil plays an active role in this world. A vintage aphorism of Gödel declares: "People owe their lives to the fact that the devil preferred to let them die slowly." (III, 213). Moreover, the devil obstructs the pet theory of Leibniz: "One of the devil's ideas for making the world incomprehensible is to destroy the pre-established harmony between world and mind."

Sins are frequently mentioned in Gödel's notes: "Through sin, one becomes indebted to the devil (III, 299)." On the other hand, you shouldn't let the devil overly impose on you: "The power of the devil is small." (Which is why "the world is not very different from how it was before the Fall (III, 321)").

Those who knew Gödel report that he was uncommonly guarded in his conversations and lectures. But in the privacy of his notes, he shows no inhibition in experimenting with ideas, chasing them up like a dog running after birds. Accordingly, many of Gödel's philosophical remarks are preceded by a "perhaps". Thus *perhaps* "Man is nothing by himself (that is, without external instruction)", *perhaps* "everything is ordered in a perfectly aesthetic way", and *perhaps* "the actual realm of human understanding is the natural numbers and only these (IV, 225)". Moreover, as Gödel advices himself, "I should *perhaps* apply the mathematical method a bit more in philosophy and the philosophical method a bit more to mathematics. (V, 198)."

There are statements with no "perhaps" about them, propositions which cannot be doubted. "The fact that I perceive a mistake implies that there is one. Indeed, if there were none, this perception would be a mistake (V, 183)." Such neat switching from one level of meaning to another occurs often in his notes. It plays an important role in Gödel's incompleteness theorems, too. "Meaning of my theorem: The mind is not dead but alive. As the mind was given to us in order to find the right way, it must be made in such a way that it is impossible to deceive it (V 183)."

A large part of the discussion of foundational problems in Gödel's philosophical notes is devoted to antinomies, be they logical, semantic or theological. Again and again, his mix of categories is startling: Thus he declares that some logical quantors and junctors are good, whereas others (such as "not", "and" or "for all") are bad (IV, 256). And after a profound, but orthodox foundational remark such as "Mathematics consists of two parts: logic and construction", one finds added "These are the male and the female principle (form and matter, resting and moving, completed and unfinished) (V, 191)." Gödel is deeply imbued with Brouwer's notion of "two-ness" as a basic principle, adding however that "without time, we likely could not perceive the pair in the first place (V, 157)." In a side-note he adds: "The mistake is that we regard the world as something "finished"(V, 166)".

If there is a motto that could be placed in front of Gödel's notebooks, it is: "Concepts bring order to the chaos of objects. But what brings order to the chaos of concepts? (V, 197)" Gödel's analysis of concepts is pursued in a highly original way, apparently without restraint, rule, or method, almost like an illustration of the principle that "Anything goes." Small wonder that Gödel never saw himself as belonging to an intellectual current of the twentieth century, particularly not as being part of logical empiricism (although he grudgingly admitted that "Positivism is the only form of materialism that can still persist in the present times (V, 177)". The notebooks reveal him as utterly alien to the prevailing *Zeitgeist*. Indeed, Gödel appears as almost other-worldly. He must have felt so himself. On several occasion, he returns to the question: "Are we an intellect, or do we have one?" The notebooks give no answer.

Full Title:

Kurt Gödel, *Philosophische Notizbücher/ Philosophical Notebooks.* Edited by Eva-Maria Engelen. Volume 3: *Maximen III/ Maxims III*, Berlin: De Gruyter 2021 (332 pp., ISBN: 978-3110753257); Volume 4: *Maximen IV/ Maxims IV*, Berlin: De Gruyter 2023 (276 pp., ISBN: 978-3110772944); Volume 5: *Maximen V/ Maxims V*, Berlin: De Gruyter 2023 (208pp., ISBN: 978-3111081144).

Chapter 16
Review: John Preston (ed.), *Interpreting Mach. Critical Essays*. Cambridge University Press 2021

Philipp Leon Bauer

Ernst Mach (1838–1916) was, no doubt, one of the most prominent scientists of the late nineteenth and early twentieth centuries. Mach, a physicist and philosopher, was a pioneer for an integrated history and philosophy of science and had a major influence on the origins and evolution of Logical Empiricism and Vienna Circle. Mach had also a major influence on the methodological development of various sciences. His doctrine of elements, historical-critical method and the principle of economy (also economy of thought, thought economy) played a crucial role in diverse scientific fields. *Interpreting Mach* focuses on Mach's epistemological and methodological ideas.

John Preston has written the introduction to the book followed by 13 critical essays on the work and thought of Ernst Mach. Preston presents an overview of Ernst Mach's life and influence as well as the presented essays (p. 1–9).

In the first paper "Ernst Mach's Piano and the Making of a Psychophysical Imaginarium" (p. 10–27) Alexandra Hui deals with Mach's work on psychophysics of accommodation in hearing. Hui concludes that "Mach's early studies of accommodation in hearing were foundational for his later historical epistemology" (p. 25). Besides discussing Mach's books and lectures, Hui's paper is enriched by other primary sources like his notebooks and correspondence with the essayist Eduard Kulke (1831–1897).

In the following paper "Mother's Milk and More: On the Role of Ernst Mach's Relational Physics in the Development of Einstein's Theory of Relativity" (p. 28–47), Richard Staley focuses on Mach's concept of physical theories regarding the development of Albert Einstein's (1879–1955) theory of relativity. Especially, Staley examines the influence of Mach's *The Science of Mechanics* (1883) on Einstein.

P. L. Bauer (✉)
Institute Vienna Circle, University of Vienna, Vienna, Austria
e-mail: philipp.leon.bauer@univie.ac.at

G. Schiemer (ed.), *The Legacy of the Vienna Circle*, Vienna Circle Institute Yearbook 30, https://doi.org/10.1007/978-3-031-80568-4_16

In "Meaningful Work: Ernst Mach on Energy Conservation" (p. 48–66) Daan Wegener claims that the law of energy conservation is highly relevant in Mach's thought. This relevance of the law is in Mach's early work, e.g., *History and Root of the Principle of the Conservation of Energy* (1872), *The Principle of the Theory of Heat* (1896) and cannot be overlooked. He points to various aspects of debates on energy conservation in the nineteenth century and explicitly addresses the varied terminologies of the law of energy conservation. S.G. Sterrett discusses Mach's conception of scientific method in "Mach on Analogy in Science" (p. 67–83). She deals with Mach's interpretation and meaning of analogy in science. According to Sterrett, Mach understood analogy as a "relation between *systems of* (*interrelated*) *concepts*" (p. 82).

The paper "Ernst Mach's Enlightenment Pragmatism: History and Economy in Scientific Cognition" (p. 84–102) illustrates Mach's relation to the American Pragmatism, with particular attention paid to Charles S. Peirce (1839–1914) and William James (1842–1910). Thomas Uebel, who also focuses on the principle of economy, concludes that "Mach's historicist naturalism can be considered an original form of pragmatism" (p. 99).

The next paper "On the Philosophical and Scientific Relationship between Ernst Mach and William James" (p. 103–122), written by Alexander Klein, focuses on the correspondence and friendly contacts between both thinkers. Klein argues that Mach's philosophical worldview developed from empiricism towards strong instrumentalism.

The common philosophical viewpoints of Ernst Mach und Friedrich Nietzsche (1844–1900) are discussed in "Ernst Mach and Friedrich Nietzsche: On the Prejudices of Scientists" (p. 123–141) where Pietro Gori argues for the similarities of their respective philosophies. For example, Mach and Nietzsche shared an anti-realistic conception of science.

Lydia Patton argues in her very interesting paper "Abstraction, Pragmatism, and History in Mach's Economy of Science" (p. 142–163) that "Mach's epistemological methods cannot be captured in their entirety without an account of his pragmatism, historicism, and abstractionism" (p. 159), whereby these interact in the economy of science.

In another article, Luca Guzzardi focuses on the well-known doctrines of Mach's epistemology. In the paper "Holding the Hand of History: Mach on the History of Science, the Analysis of Sensations, and the Economy of Thought" (p. 164–183), Guzzardi deals in particular with the relation between the principle of thought economy and the historical-critical method.

In "Ernst Mach and the Vienna Circle: A Re-evaluation of the Reception and Influence of His Work" (p. 184–207) Friedrich Stadler shows Mach's major influence, especially with his principle of economy, on the so-called "First Vienna Circle",[1] which was an informal discussion group before World War I, mainly

[1]About the term "First Vienna Circle": Haller, Rudolf. 1985. "Der erste Wiener Kreis". In *Erkenntnis* (Vol. 22), 341–358. Dordrecht: D. Reidel Publishing Company.

Criticism of the term:

consisting of Philipp Frank (1884–1966), Otto Neurath (1882–1945), and Hans Hahn (1879–1934). Stadler goes on to outline the influence on the later Vienna Circle. Moritz Schlick (1882–1936), Rudolf Carnap (1891–1970), Karl Menger (1902–1985) and Richard von Mises (1883–1953) are considered in more detail. Mach's influence can also be found in the so-called "Kraft Circle", which emerged around Viktor Kraft (1880–1975), after World War II. In this context, Stadler mentions Paul Feyerabend (1924–1994) who emphasized Mach in his work.

In "Narratives Divided: The Austrian and the German Mach" (p. 208–234) Michael Stöltzner compares the reception of Mach in Austria with Mach's very different reception in Germany. Stöltzner, who makes exciting references to physicists, e.g., Albert Einstein (1879–1955), Ludwig Boltzmann (1844–1906), and Max Planck (1858–1947), provides a very good overview of his views at the end of his article.

The editor of this volume, John Preston, presents his own essay titled: "Phenomenalism, or Neutral Monism, in Mach's *Analysis of Sensations*?" (p. 235–257). For Preston, Mach was not a phenomenalist and he argues that the book *The Analysis of Sensations* (1886) can be considered a book espousing neutral monism.

The book concludes with the paper "The Case for Mach's Neutral Monism" (p. 258–279) by Erik C. Banks (1970–2017). Banks, who unfortunately died too early, argues against the point of view that Mach was a phenomenalist and takes a neutral monism position in this context. He also compares the position of Mach with later elaborations of neutral monism by William James and Bertrand Russell (1872–1970).

John Preston's collection of papers in this volume constitute a positive contribution to the available literature. Due to the wide range of topics (p. i), the book is aimed at a broader readership. This book is highly appealing to researchers of various sciences and for those who are interested in Mach in particular. However, despite the wide range of topics presented in this book, many papers are thematically related to one another, and therefore connections can be made among the papers in the book.

This book further represents an important contribution to the current international interest in Ernst Mach, e.g., *Mach: Life, Work, Influence* ed. by Friedrich Stadler (2019). Mach's revival is at its beginning and will continue to provide an exciting field of research for many sciences. To summarize, *Interpreting Mach* is an exciting collection of critical essays now available to Mach scholars as well as others interested in Mach's innovative thought.

Full Title:
John Preston (ed.), *Interpreting Mach. Critical Essays*. Cambridge: Cambridge University Press 2021. pp. xii + 287, ISBN: 978-1-108-47401-6.

Limbeck-Lilienau, Christoph. 2022. "The First Vienna Circle and the Erlangen Conference". In *The Routledge Handbook of Logical Empiricism*, ed. by Thomas Uebel and Christoph Limbeck-Lilienau, 99–108. London/New York: Routledge.

Index

G. Schiemer (ed.), *The Legacy of the Vienna Circle*, Vienna Circle Institute Yearbook 30, https://doi.org/10.1007/978-3-031-80568-4

The manufacturer's authorised representative in the EU is Springer Nature Customer Service Centre GmbH, Europaplatz 3, 69115 Heidelberg, Germany. If you have any concerns regarding our products, please contact ProductSafety@springernature.com

Printed and bound by CPI Group (UK) Ltd, Croydon, CR0 4YY
01/10/2026
02225950-0007